LIFE and DEATH
in an Oxygen Atmosphere

Manfred K. Eberhardt

BookSurge LLC

2007

Published by BookSurge LLC
An Amazon.com company
7290 B Investment Drive
Charleston, SC 29418
Website: HYPERLINK "http://www.booksurge.com"
www.booksurge.com
ISBN: 1-4196-6693-2
Available from: HYPERLINK "http://www.booksurge.com"
www.booksurge.com
HYPERLINK "http://www.amazon.com"
www.amazon.com
HYPERLINK "http://www.borders.com"
www.borders.com

Acknowledgements

I wish to thank Ms. Myrna Cabán and Mrs. Aimee Berolo for their help in preparing all the formulas and putting the entire manuscript in camera-ready form. I also wish to thank Mr. Luis Deyá for his contribution to the cover design.

Manfred K. Eberhardt   San Juan, 2004.

Front Cover: "Mirror universes (1996)" sculpture by the author. Laminated Formica, 24" x 16" x 16". Cover design by Myrna Cabán.

TABLE OF CONTENT

INTRODUCTION

Oxygen has long been recognized as a life-sustaining molecule. Another view, attesting to its now well-recognized lethality, is that oxygen is the worst environmental pollutant of all time.

Linda Biadasz Clerch
Donald J. Massaro

Everybody is talking about antioxidants. Newspapers, magazines and television report the beneficial effects of antioxidants, and the pharmaceutical industry is selling billions of dollars worth of these dietary supplements, but very few people really know what antioxidants are, why they exist and how they do what they do. Oxygen plays a central role in these reactions. People usually believe that oxygen is something good for our health and they are completely unaware of the numerous damaging and even lethal effects of oxygen. This state of blissful ignorance has many causes. The reactions of oxygen and antioxidants are part of chemistry. Chemists have not been as successful in making their science accessible to the general public as biologists or medical professionals. Chemistry in the public mind has a negative image, associated with poisons, pollution, environmental degradation and cancer causing chemicals. Another reason for this state of affairs lies in the subject matter itself. Chemistry is one of the basic, hard sciences, which has developed its own language based on formulas and equations, whereas biology and medicine are more descriptive. Words are, of course, never as precise as formulas and equations. The disdain for words was very well expressed by Mephisto in Goethe's Faust:

2

All right. But do not plague yourself too anxiously:
For just where no ideas are, the proper word is
never far.[*]

On the other hand, words are easier understood by the general public. Whenever somebody asks me what I do for a living and I say "I am a chemist', they throw up their arms and tell me that chemistry was their worst subject in school. In chemistry courses we learn innumerable facts, which have to be memorized , rather than understood. Everything in science is simple at the most basic level. Many details may appear confusing and many controversies and **paradoxes** arose due to an incomplete understanding of the basic chemistry involved. Once we understand the basic concepts, everything else falls into place. Biology and medicine are reduced to the chemistry of deoxyribonucleic acid (DNA), ribonucleic acid (RNA) and proteins.

As a professor in the Department of Pathology in a Medical School, I was, of course, aware of the importance of oxygen and reactive oxygen metabolites (ROMs) to the medical profession. I have therefore written a book on the subject (CRC Press, 2000). Several professors at the Medical School told me, that the book was too difficult for medical students, residents and students in the biomedical sciences. I hope to remedy this situation with the present book, in which I try to translate the chemistry of oxygen metabolites into words and try to keep formulas and equations to a minimum. To teach chemistry only with words is mission impossible. It is like trying to communicate with a Frenchman using only your hands and feet. I assume, therefore, that the readers are somewhat scientifically literate. Unfortunately, even science majors, taking fresh-

---

[*] The quotation from Goethe's Faust is from the translation by Walter Kaufmann, Anchor Books, Doubleday, New York, 1961.

man organic chemistry, get little or no exposure to the chemistry of radicals. **Radicals** are highly reactive species and for many decades were considered of minor significance in biological systems. However, the oxygen metabolism was found to lead to **reactive oxygen metabolites,** some of which are radicals. ROMs play an important role in many normal biological as well as pathological processes. At present there is evidence for radical involvement in over 100 diseases. We no longer have to concern ourselves only with outside agents (bacteria, viruses, pollution) that affect our health, but also with damage caused by agents of normal oxygen metabolism.

Fortunately our bodies are not defenseless against the damaging effects of reactive oxygen metabolites. During the course of evolution, our bodies have developed **defenses against ROMs**. These defenses consist of the enzymes superoxide dismutase (SOD), catalase (CAT), glutathione peroxidase (GSH-Px), metal complexing proteins, uric acid, melatonin and small molecules taken up in our diet, the vitamins A, C and E.

The basis of biology is chemistry. Once we understand the chemistry of ROMs we can better rationalize the pathology. Numerous diseases with diverse clinical symptoms can be understood at the most basic level by the concept of **'oxidative stress'**, which is a disturbance of the **prooxidant/antioxidant balance** of a system. The importance of balance has been recognized for centuries by Chinese philosophers (Yin-Yang) and was expressed by Paracelsus: everything is poison, it just depends on the dose. The Yin-Yang symbol is shown on the back cover. The two dots in the symbol represents the idea that each time one of the two opposing forces reaches the extreme, it contains in itself already the seeds of its opposite.

The front cover of this book shows a sculpture by the author representing two mirror images of a mathematical construct known as a Möbius strip. You may ask: what does this sculpture have to do with reactive oxygen metabolites. The sculpture has two sides, but only one surface. Everything in life has two sides, pro or con, good or evil, body or spirit, black or white, day or night, life or death. One side cannot exist without the other. In life there is death.

The thread of pro and con runs through most of radical biology and medicine. Our cells are carrying out a precarious balancing act like the "Fiddler on the Roof". The concentrations of many chemicals in our bodies, such as Fe(III)/Fe(II), Cu(II)/Cu(I), O_2, H^+, Ca^{2+}, have to be strictly controlled. The **antioxidants** ascorbic acid (Vitamin C) and α-tocopherol (Vitamin E) are essential for life, but under certain conditions can have damaging (**prooxidant**) effects. The prooxidant and antioxidant effects of Vitamin C and E have been extensively investigated and will be discussed in detail in Chapter 4. The possible role of Vitamin C as a slow acting carcinogen has been debated (Halliwell, 1994). We encounter this pro and con situation also with other antioxidant defenses, produced by our own bodies. One of these antioxidants is the enzyme **superoxide dismutase (SOD)**. Too much or too little of this antioxidant has damaging consequences. It is clearly a question of balance. During the course of this book I shall discuss numerous examples of pros and cons in reactive oxygen metabolite chemistry and pathology.

The reactive oxygen metabolites are damaging to our cells and the accumulated damage leads to aging, cancer and other degenerative diseases. Thus it appears that death is inherent with life and programmed already from the very beginning through the oxygen metabolism. The above statement by Clerch and Massaro expresses the dual character of oxygen as symbolized by the sculpture.

Chapter 1 introduces some basic chemical and biological concepts and definitions, such as the nature of the chemical bond, radicals, reactive oxygen metabolites, stereochemistry, electron transfer, homeostasis, hydrogen bonding, singlet oxygen, electronically excited states, cell structure, metabolism and reproduction. I start with the basic building blocks of life, and discuss briefly the evolution of life from these chemicals and the arrival of oxygen. The appearance of oxygen represents the most important event in the evolution of life on our planet. It is accompanied with the evolution of more complex cells (eukaryotes), multicellular organisms and the appearance of death. Unicellular organisms (the blue-green algae) were immortal. Death is partly a consequence of complexity and oxygen metabolism.

This text is on an interdisciplinary subject, involving chemistry, biology and medicine. What may be obvious to a chemist may be difficult to grasp for a biologist or a medical student or vice versa. It is therefore unavoidable to discuss some topics, which are already quite familiar to some readers.

In order to understand biology and medicine we have to understand the underlying chemistry. An important part of this book is therefore devoted to the chemistry of oxygen and its metabolites.

Chapter 1.

THE CHEMISTRY OF LIFE

The living state is the electronically desaturated state of protein. When life originated three and a half billion years ago, our globe was covered by a dense layer of water vapor. There was no light and no oxygen near its surface.
Albert Szent Györgyi in 'The living state and cancer'

The basic building blocks.

Before life can start, we need the essential building blocks to construct life (chemical evolution). Since life is based on carbon-containing compounds in aqueous solutions, we require certain strict conditions, which have to be met before life can evolve. The temperature has to be between 0°C and 100°C (the melting and boiling point of water). In our Universe, temperatures range from -273°C (absolute zero) to over 1 million °C. We can see that we indeed occupy a very narrow niche in the Universe.

The periodic table of the elements consists of 92 elements (plus a series of artificially created ones). However the major part of living systems consists of only six of them, namely carbon (C), hydrogen (H), oxygen (O), nitrogen (N), phosphor (P) and sulfur (S) as important elements. The carbon atom occupies a special place in the periodic table (see appendix). It is in the second row of the table and is in the middle. All elements to the left (lithium, beryllium and boron) are less electronegative than carbon and all the elements to its right (nitrogen, oxygen and fluorine) are more electronegative. A stable chemical bond consist of electrons, which always come in pairs (two, four or six electrons):

$$-\overset{|}{\underset{|}{C}}-\overset{|}{\underset{|}{C}}- \qquad \overset{\diagdown}{\diagup}C=C\overset{\diagup}{\diagdown} \qquad -C\equiv C-$$

2 e⁻ bond	4 e⁻ bond	6 e⁻ bond
(single bond)	(double bond)	(triple bond)

The carbon atom can form a total of four bonds, whereas other atoms (nitrogen or oxygen or hydrogen) can only form three, two or one bond respectively:

$$N\equiv C- \qquad O=C\overset{\diagup}{\diagdown} \qquad H-\overset{|}{\underset{|}{C}}-$$

The possibilties of building different molecules from carbon atoms are therefore infinitely greater than with other atoms.The great variety of life depends on this basic fact of chemistry. We can build up long carbon-carbon chains as follows:

$$-\overset{|}{\underset{|}{C}}-\overset{|}{\underset{|}{C}}-\overset{|}{\underset{|}{C}}-\overset{|}{\underset{|}{C}}-\overset{|}{\underset{|}{C}}-\overset{|}{\underset{|}{C}}- \quad \text{or} \quad -\overset{|}{\underset{|}{C}}-\overset{|}{\underset{C}{C}}-\overset{|}{\underset{|}{C}}-\overset{|}{\underset{|}{C}}-\overset{C}{\underset{C}{C}}-\overset{|}{\underset{|}{C}}-$$

This class of compounds is known as hydrocarbons. These hydrocarbons are all known to us as constituents of crude oil. The long chain hydrocarbons are converted to lower molecular weight hydrocarbons via the process of hydrocarbon cracking:

$$-\overset{|}{\underset{|}{C}}-\overset{|}{\underset{|}{C}}-\overset{|}{\underset{|}{C}}-\overset{|}{\underset{|}{C}}-\overset{|}{\underset{|}{C}}-\overset{|}{\underset{|}{C}}- \xrightarrow{\text{heat}} -\overset{|}{\underset{|}{C}}-\overset{|}{\underset{|}{C}}-\overset{|}{\underset{|}{C}}\cdot + \cdot\overset{|}{\underset{|}{C}}-\overset{|}{\underset{|}{C}}-\overset{|}{\underset{|}{C}}- \quad (1)$$

$$\downarrow$$

$$-\overset{|}{\underset{|}{C}}-\overset{|}{\underset{|}{C}}-\overset{|}{\underset{|}{C}}- + \overset{\diagdown}{\diagup}C=\overset{|}{\underset{|}{C}}-\overset{|}{\underset{|}{C}}-$$

In this way we transform the long chain and high boiling hydrocarbons to smaller hydrocarbons with a lower boiling point. This is essential for use as a fuel in internal combustion engines. The initially formed fragments in reaction (1) are known as radicals. **Radicals** have one unpaired electron as indicated by the dot.

The simplest hydrocarbon is methane (CH_4):

$$
\begin{array}{c}
H \\
| \\
H\text{-}C\text{-}H \\
| \\
H
\end{array}
$$

The hydrogen atoms can be replaced by C, O, N and other elements (like P or S) and in this fashion we can schematically build up all the basic building blocks of life. These molecules are the sugars, amino acids, fatty acids, pyrimidines and purines (see Tables I and II, appendix).

From these building blocks we can build up the construction material for living forms, namely carbohydrates, proteins, fats, **deoxyribose nucleic acid (DNA)** and **ribose nucleic acid (RNA).** Carbohydrates are sugars linked together in long chains, proteins are chains of amino acids and fats are fatty acids linked to some alcohols. Proteins come in two varieties: structural proteins, used for the construction of cells, and enzymes, the catalysts of biochemical reactions. Cellulose, the building material of trees is a polymerized glucose and is insoluble in water, but can be split into glucose via enzymatic digestion. Some of these enzymes are present in our saliva. Sugar and sawdust are basically the same, despite their different appearance and characteristics.

Proteins contain hundreds or thousands of amino acids linked together as follows:

$$R\text{-}CH\text{-}COOH \quad + \quad R'\text{-}CH\text{-}COOH$$
$$\underset{NH_2}{|} \qquad\qquad \underset{NH_2}{|}$$

$$\downarrow$$

$$\overset{R'}{\underset{\;}{|}}$$
$$R\text{-}CH\text{-}CONH\text{-}CH\text{-}COOH \quad + \quad H_2O$$
$$\underset{NH_2}{|}$$

This process just involves the elimination of water (H_2O) and can be repeated many times, and is called copolymerization (Fox, 1988). Amino acids connected in this way are called peptides or polypeptides. Depending on the structure of R and R' the copolymerization can be accomplished in a test tube (*in vitro*) by simply heating the mixture (Fox, 1988). *In vivo,* the copolymerization is accomplished by enzymes. Enzymes (the biochemical catalysts) are not just long chains of connected amino acids, but they have a three-dimensional structure. This structure is determined among other factors, by hydrogen bonds between individual amino acid groups. The three-dimensional structure is essential for the enzymes biochemical activity. The DNA and RNA are somewhat more complex. DNA and RNA consist of a sugar-phosphate backbone to which are attached the bases, thymine (or uracil), cytosine, adenine and guanine. The difference between RNA and DNA is the sugar, which is ribose in RNA and deoxyribose in DNA, and uracil replaces thymine in RNA. Uracil is distinguished from thymine by the absence of a CH_3 group. In non-dividing cells the DNA is a **double-stranded helix** (like two strands of a rope). A schematic representation of a single strand of DNA is given in Table III, appendix. The phosphate group ($-PO_3H$) is at neutral pH dissociated to $-PO_3^-$ and H^+. All DNA on earth forms a right-handed spiral.

The individual units consisting of base-sugar-phosphate are called nucleotides. DNA and RNA are high molecular weight polymers consisting of many nucleotides. In the double stranded DNA the two strands are held together by hydrogen bonds between the bases. These hydrogen bonds are specific between adenine-thymine (or uracil in RNA) and guanine-cytosine (Table III, appendix). These base pairs are complementary.

The four bases (A, G, C, T, Table II, appendix) specify all the instruction for the build up and the proper functioning of an organism. Combinations of three bases specify one of the 20 amino acids, which make up all the proteins and enzymes of living forms. The amino acids are summarized in Table I (appendix).

DNA is very important for the survival of the species and is well protected from outside influences. First DNA exists as a double strand and this double strand is surrounded by histones. Histones are small molecular weight basic proteins (containing arginine and lysine). The histones at neutral pH are positively charged and are located in the deep groove of the DNA double helix. The positive charges on the histones form bonds with the negatively charged phosphate groups of DNA. The DNA in its resting state is therefore well protected against outside influences. The DNA is however vulnerable whenever the cell divides and the DNA is present as a single strand. Cell division (mitosis) is therefore important for DNA damage. A link between **mitosis and carcinogenesis** has indeed been established. I shall come back to this topic again in Chapter 5.

The DNA of each individual human is distinct, but the architecture (the double-helical structure) is the same in all organisms such as bacteria, flies, mice and humans. During the whole of evolution the basic chemistry of all living forms has not changed. Different animals and humans use the same amino acids to synthesize their proteins. The same chlorophyll is used in all green plants and the same hemoglobin is used in animals and humans to transport oxygen to different tissues.

The basic molecules of life as shown in Tables I-III (appendix) are, of course only two-dimensional representations of three-dimensional structures. A molecule of methane (CH_4) has the form of a tetrahedron. The four H-atoms occupy the four corners of the tetrahedron and the C-atom is at the center:

$$
\begin{array}{c}
H \\
| \\
C \\
H \diagup \ \ \diagdown H \\
H
\end{array}
$$

The tetrahedral structure of methane appears quite obvious in hindsight, but the idea, proposed by van't Hoff and LeBel, was vehemently opposed by the scientific establishment. So what else is new? The attacks by the famous German professor Hermann Kolbe were particularly nasty. In 1901 **van't Hoff** was the first chemist to receive the Nobel Prize for Chemistry.

Whenever we have a C-atom with four different substituents, we can have two stereo arrangement, which are mirror images of each other.

$$
\begin{array}{ccc}
B & & B \\
| & & | \\
A\text{-}C\text{-}C & & C\text{-}C\text{-}A \\
| & & | \\
D & & D
\end{array}
$$

In one case we have the four substituents, A, B, C, and D arranged in a clockwise fashion and in the mirror image they are arranged counter clockwise. These two isomers are like our two hands and are therefore called left-handed or right-handed. They cannot be superimposed. These mirror images are 'optically active', i.e. they rotate the plane of polarized light in opposite directions. Amino acids (appendix) have one asymmetric center (indicated by a *) and they therefore exist in two 'enantiomeric forms' (a L and a R form). When we synthesize amino acids in a test tube, we always produce equal amounts of both enantiomers. However in living systems only one (the L-form) is produced and used for the construction of proteins. We encounter this selectivity in all living forms. **Louis Pasteur**, who first discovered this phenomenon suggested that the presence of only one enantiomeric form in living forms distinguishes them from non-living matter. Surprisingly, no species has yet evolved, which uses the other (the R) form for the synthesis of its proteins. Why *Nature* uses only one form has, until recently, been one of the unsolved mysteries of Science. In 1997 a group of Japanese researchers showed that polarized light in outer space created a small preponderance of L-amino acids. Subsequently chemical reactions amplified this small excess and drove the right handed aminoacids towards extinction (see John L. Casti, 2000).

Aristotle proposed that a special life force was necessary to explain the remarkable properties of living organisms. Chemists believed for a long time that organic molecules could not be synthesized in a test tube, but only by living creatures This belief has become known as **vitalism**.

The rejection of vitalism has usually been traced back to the synthesis of urea from ammonium isocyanate (NH_4CNO), an inorganic compound by **Friedrich Wöhler** (1828). However, as pointed out by D. McKie (1944) Wöhler's synthesis did not convince too many people of the unnecessary life force for the synthesis of biological compounds. It took the work of several other chemists, including the synthesis of acetic acid by **Hermann Kolbe** in 1845, to convince people. As stated by McKie: "vitalism in organic chemistry was rejected not by any sudden and dramatic synthesis - for science does not advance and *Nature* does not reveal herself in that way - but by steady accumulation of contradictory facts". Science is a continually evolving structure.

   Another form of stereo isomers is found in unsaturated fatty acids. Whenever we have a C=C double bond with two substituents, they can be either on the same side of the double bond (cis) or on opposite sides (trans):

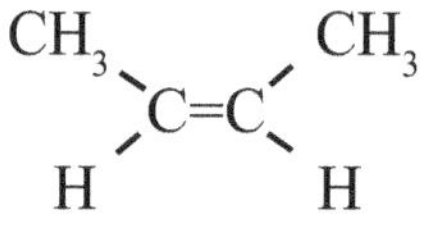

$$CH_3 \diagdown \diagup CH_3$$
$$C=C$$
$$H \diagup \diagdown H$$
cis -butene

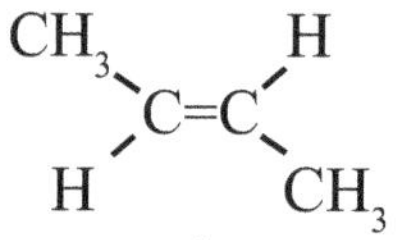

$$CH_3 \diagdown \diagup H$$
$$C=C$$
$$H \diagup \diagdown CH_3$$
trans-butene

Unsaturated fatty acids are essential for reproduction and growth and have to be obtained through the diet. These fatty acids cannot be synthesized by our own bodies and are called "**essential fatty acids**" (see p.29). Plants are especially rich in linoleic acid. Another typical unsaturated fatty acid is oleic acid. The structure of these two molecules are as follows:

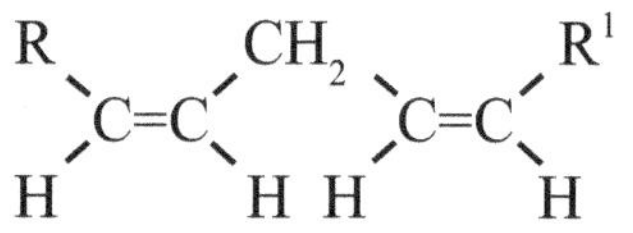

$$R \diagdown \diagup CH_2 \diagdown \diagup R^1$$
$$C=C \qquad C=C$$
$$H \diagup \diagdown H \ H \diagup \diagdown H$$
cis, cis-linoleic acid

where R is -$(CH_2)_4$-CH_3 and R' is - $(CH_2)_7$-COOH

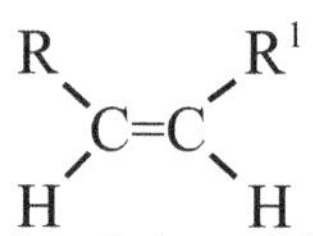

$$R \diagdown \diagup R^1$$
$$C=C$$
$$H \diagup \diagdown H$$
cis-oleic acid

where R is -$(CH_2)_7$-CH_3 and R' is - $(CH_2)_7$-COOH

Fatty acids with more than one double bond are known as polyunsaturated fatty acids (PUFAs). Most naturally occuring unsaturated fatty acids are in the cis configuration. Only the cis forms are used by animals and humans for essential functions. However, in recent history *trans* **fatty acids** have been introduced into our food supply by partial hydrogenation of polyunsaturated fatty acids. It has been determined, that *trans* fatty acids in our diet are responsible (in part) for the increased development of **coronary artery disease** (Willett et al. 1993, see under atherosclerosis). The more unsaturated fatty acids a fat contains, the more liquid it is (candles, margarine, butter, olive oil). Our bodies need both saturated, unsaturated and polyunsaturated fatty acids. The different properties of the fatty acids offers *Nature* a wide variety of building materials for different structural requirements. For example, the outer layer of a neuron, the myelin sheath, uses saturated fats to construct a rigid hull. On the other hand, the inner lining of the arteries and veins, which must be flexible, contain high levels of unsaturated fatty acids.

The geometry of molecules is very important for their biological functions. The importance of geometry was already recognized by the Pythagorean philosophers, who believed that geometry is the key to the Universe. There are numerous examples of chiral drugs, in which only one enantiomeric form is active. This fact has been duly noted by the pharmaceutical industry, and the number of chiral drugs consisting of only one enantiomer has been steadily increasing during the last decades. Although the synthesis of asymmetric molecules is the normal pathway in biological systems, it is no mean feat in a test tube. The 2001 Nobel Prize in Chemistry was awarded to three scientists (**William S. Knowles, Ryoji Noyori and K. Barry Sharpless**) for the development of procedures for the synthesis of chiral molecules. According to a recent report in

Chemical and Engineering News (Oct. 23, 2000) the sale of chiral drugs has reached over $ 100 billion.

Thalidomide has become an infamous drug during the 1950s for causing severe birth defects if taken by pregnant women. Thalidomide is a chiral molecule, i.e. it exists in two forms (left and right handed). It has been claimed that in the case of thalidomide only one enantiomeric form is teratogenic. Unfortunately, the non-teratogenic form is converted to the teratogenic form under physiological conditions. Recently thalidomide has made a comeback as an anticancer drug (see chapter 5, p.129).

Evolution of complex molecules and cells.

The search for extraterrestrial life has rekindled the interest in the origin of life. In order to determine how life originated we first have to ask the questions: what is life? Can we explain the origin of life by purely physical and chemical processes?

The important characteristics of life are: **metabolism and reproduction.** Metabolism requires enzymes (protein) and reproduction requires polynucleotides (DNA). The mathematician **John v. Neuman** (1948) first described an analogy between the functioning of living organisms and the functioning of a computer. A cell consists of hardware (membranes, proteins) and software (DNA). Enzymes (protein) are essential for metabolism and DNA is essential for reproduction. Organisms, which only have software (polynucleotides) cannot exist independently, but must be obligatory parasites, they use the hardware (metabolism) of the host. In order for the computer to work it needs energy (see Brown, 2000). The power plant of the cell are the mitochondria. Mitochondria produce energy via oxidative metabolism and store this energy in the form of a high energy chemical molecule, known as adenosinetriphosphate (ATP). ATP is the battery and is the essential catalyst for many cellular functions.

Any damage to the hardware (lipids, proteins) or software (DNA/RNA) or to the power plant (mitochondria) will impair cellular function. All of these molecules are damaged by reactive oxygen metabolites (ROMs).

Concerning the question of how life originated, we are faced with the following problem: what came first, the hardware (proteins) or the software (polynucleotides)? DNA contains all the information for the synthesis of enzymes (proteins), and enzymes are essential for the synthesis of DNA/RNA from the monomeric nucleotides. We have a chicken or egg problem. A detailed discussion of the different origin of life theories can be found in some recent books by F. Dyson, P. Davies and J. L. Casti.

For a long time it was thought that any organism, which duplicates must contain polynucleotides. Proteins cannot duplicate and DNA or RNA cannot catalyze (fulfill enzymatic functions). However, there are some exceptions to these pillars of molecular biology. The progress of science is to search for these exceptions and thus develop new theories. This is the way science advances. As expressed by John L. Casti - solving the Great Mysteries of science is akin to peeling the skin of an onion. We get closer and closer to the core - but never quite reach it. According to these points of view there never can be a final theory. For a contrary view see Steven Weinberg in "Dreams of a Final Theory". The discovery of exceptions is usually rewarded with a Nobel Prize. **Carlton Gajdusek** received the Prize in 1977 for his work on kuru (caused by an abnormal protein named amyloid), **Stanley Prusiner** in 1999 for his work on prions (proteinaceous infectious particles), which cause scrapie in animals, Creutzfeldt-Jacob and

Gershmann-Sträussler-Scheinker disease in humans, and **Thomas Cech and Sidney Altman** in 1989 for their discovery of the catalytic effect of RNA. The discoveries of Cech and Altmann makes an RNA world possible. In this scenario neither proteins nor DNA came first, but RNA (Gilbert, 1986, Schimmel and Alexander, 1998). Contrary views have been expressed (Joyce, 1989) and the search is on for the pre-RNA world (Piccirilli, 1995)

On the other hand, according to **Lynn Margulis** the original living organisms were cells with only a metabolic apparatus, directed by enzymes, but without polynucleotides (genetic apparatus). The RNA first appeared as a parasite within the cell. Slowly over millions of years the parasite became a symbiont, producing a cell with both hardware (metabolism) and software (replication). In addition Margulis collected evidence which showed that the mitochondria once were independent organisms, which migrated into the eukaryotic cells and lived there as parasites. These primitive cells slowly adapted to these parasites and formed a symbiotic relationship. The fact that mitochondria have their own DNA supports this hypothesis.

The first life forms, the blue-green algae contained DNA, but no nuclei, and they did not die. They just kept on dividing forever and stayed around for about 1.6 billion years before going into decline. The more complex cells of today contain nuclei (DNA surrounded by a membrane) and many sub-cellular units, each performing specialized functions. In order to make this possible these organelles are surrounded by membranes. These organisms are known as eukaryotes as opposed to prokaryotes without inner membranes. The organization of different functions into different locations and the membrane surrounding the organelles is the characteristic of eukaryotes and distinguishes them from the more primitive bacteria and blue-green algae. The blue-green algae lived in an **oxygen free atmosphere**, and they produced oxygen from carbon dioxide and water, using photo energy (photosynthesis). However it took a long time to build up enough free oxygen in the atmosphere for aerobic (oxygen consuming) organisms to evolve. There are lots of elements around in the earth's crust which thirst for oxygen and react rapidly (sometimes violently) with oxygen to produce oxides, such as iron oxides, copper oxides, manganese oxides, chromium oxides etc. etc. Free oxygen only became available after all the metals had quenched their thirst for oxygen. Only then did more complex eukaryotic cells evolve. The absence of free oxygen in the atmosphere was the reason why the blue-green algae dominated the biosphere for so long.

The evolution of multicellular organisms of course also requires membranes to separate one cell from the other. Most membranes contain polyunsaturated fatty acids (PUFAs), which are easily damaged by oxygen This process is known to all of us as **rancification**. The evolution of PUFAs is essential, since the unsaturation of these molecules makes the membrane more flexible. The unsaturation, which makes the PUFAs a useful material for the construction of membranes, makes it also a potential target for damage by ROMs (pro and con situation). Hand in hand with increasing complexity came specialization. Cells organized into organs. This division of labor makes the machinery run more efficiently. Let's look at a schematic and simplified picture of a cell (Fig. 1).

A human body consists of about 100 trillion cells, all of which fulfill specific functions. A cell consists of a nucleus (containing DNA), the cytoplasm (consisting of proteins, enzymes and water) and a membrane (lipids). Within the cytoplasm there are smaller structures (organelles), which in eukaryotic cells are all surrounded by a membrane. Some important organalles are the mitochondria, ribosomes and the lysosomes. The ribosomes and lysosomes fulfill opposite functions. The ribosomes are the manufacturing plants of the cell. In the ribosomes new proteins are synthesized according to the instructions contained in the DNA. The lysosomes, on the other hand, contain enzymes, which take damaged proteins apart to their amino acid subunits, which then are reused for the manufacture of new proteins (recycling). The cell is constantly building new molecules and tearing old ones apart.

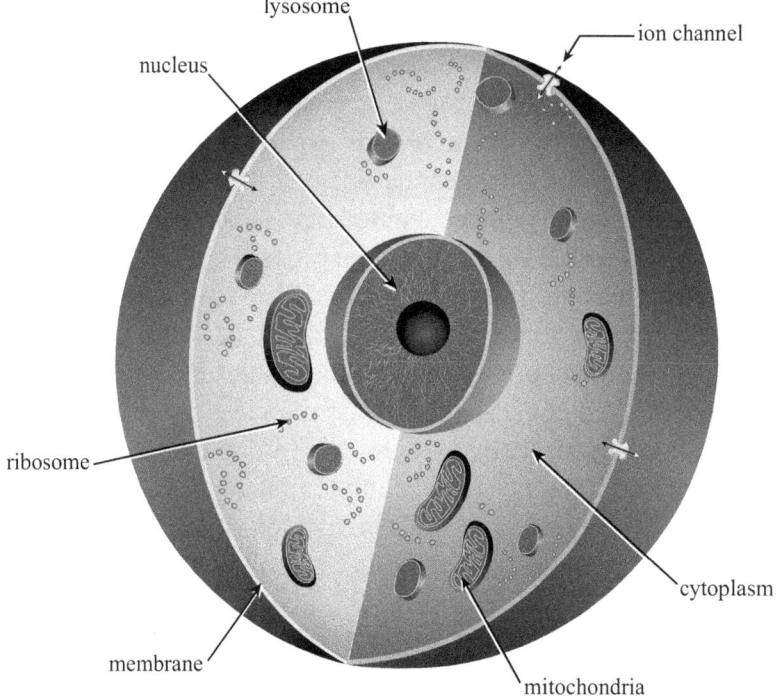

Fig. 1. Eukaryotic cell.

For the program to run the computer needs energy, which is supplied by the mitochondrion, the power plant of the body (see Brown, 2000). The energy is obtained through **oxidation reactions**. Oxidation is the removal of one or more electrons. The electrons of course have to go somewhere, they react with another molecule, which is reduced. In other words **oxidation and reduction** are two interdependent processes, one cannot exist without the other. Translated into ordinary language this would be: to give and to take are interdependent, as was already recognized by the wise men who wrote the Bible: The Lord giveth and the Lord taketh away. On the other hand we have the paradoxical statement: to give is holier than to receive. **Paradoxes** are obviously as old as humankind. For a discussion of some paradoxes, see D. Hofstadter (1979)**.**

The foodstuffs are oxidized (electrons are removed) via the mitochondrial respiratory electron transport chain in a series of steps involving many enzymes. The final electron acceptor in this sequence is oxygen, which is reduced in the last step by four electrons to give two H_2O (see equations below). The energy gained in this series of electron transfers is stored in the form of chemical cnergy. This is the process of oxidative phosphorylation, which forms adenosine triphosphate (ATP) from adenosine diphosphate (ADP). ATP is used as an energy source for cellular function. Whenever one of the three phospate groups in ATP is released it also releases energy, which is used for cellular functions. Without oxygen life on earth as we know it today would not exist. The atmosphere we breath consists of 21% of oxygen. There is a very limited range of oxygen concentration in which we humans can exist. Lowering or increasing the oxygen concentration by just a few percent causes damage. I shall discuss the damaging effects of oxygen in Chapter 5 (hyperoxia). These oxygen reductions proceed via one, two or four - electron transfer reactions:

$$O_2 \; + \; e_{aq}^- \; \rightarrow \; O_2^{\cdot -} \qquad \textbf{Superoxide radical anion}$$

$$O_2 \; + \; 2\,e_{aq}^- \; + 2\,H^+ \; \rightarrow \; H_2O_2 \quad \textbf{Hydrogen peroxide}$$

$$O_2 \; + 4\,e_{aq}^- + \; 4\,H^+ \; \rightarrow \; 2\,H_2O \qquad \textbf{Water}$$

The main part of the consumed oxygen (about 95%) is reduced to water, which of course is not toxic. However between 1-5% is reduced to $O_2^{\cdot -}$ and H_2O_2, which are known as **reactive oxygen metabolites (ROMs).**

We may compare the formation of ROMs with a conventional power plant or an internal combustion engine. The combustion of petroleum or gasoline produces power, but also some toxic side products, such as sulfur dioxide, nitrogen oxides, ozone, carbon monoxide, just to name a few. These toxic side products can be removed by scrubbers or catalytic converters. The catalytic converter in cells are the antioxidants, some of which are enzymes such as superoxide dismutase (SOD), catalase (CAT), glutathione peroxidase (GSH-Px), some low molecular weight molecules, such as melatonin, uric acid as well as compounds taken up in our diet, the vitamins A, C. and E.

We can damage the computer by damage to the hardware, the software or by knocking out the power plant. The theory of aging originally proposed by Harman (1981) assumed aging to be due to continuous accumulation of nuclear DNA damage caused by ROMs during the life span of the organism. Recently, evidence has emerged which seems to show that damage to the power plant (mitochondria) is the determining factor in aging (Yakes and van Houten 1997).

Evolution of scientific theories follow the same path as the evolution of species. A theory is constantly refined or changed, depending on the available evidence. In the end only the theory, which is consistent with all (or most) of the experimental data survives (survival of the fittest). As stated by J. D. Barrow, *"science is really founded on observations rather than upon facts, and so is a continually evolving structure"*. This property of change gives science its vitality. The final theory of everything may not be such a desirable goal. However a final theory would not spell the end of scientific research. As stated by S. Weinberg:" A final theory will be final only in one sense - it will bring to an end a certain kind of science, the ancient search for those principles that cannot be explained in terms of deeper principles".

Metabolism

Chemistry is the transformation of one molecule into another via reshuffling of electrons. In other words chemistry is the science of transformation and change. Taoist philosophers recognized that transformation and change are essential features of nature, and thus of life.

Chemical reactions carried out in a test tube are quite distinct from reactions occurring in cells. As any organic chemist will tell you, the synthesis of an organic molecule in the laboratory starts with a big pot containing kgs. of starting material and after several reaction sequences you wind up with a few grams or milligrams of the desired molecule. This is of course a great waste of energy , which living organisms can hardly afford. If a reaction does not proceed by itself, chemist usually heat the mixture for extended periods of time or add some catalyst. Since in a biological system we are limited to a specific temperature, biological systems use catalysts. These catalysts are the enzymes (proteins), each of which catalyzes only one specific reaction. This specificity allows a cell to carry out many different reactions simultaneously. Without this specificity we would have chemical chaos.

Enzymes are synthesized by our own cells according to the instructions encoded in DNA and are capable of greatly accelerating the rate of a specific reaction, and make a specific product in 100 % yield. How do enzymes accomplish this remarkable feat? The important part of the enzyme is its three-dimensional geometry. It contains a region known as the active site. The reactants fit into this active site like a key in a lock. Remember the Pythagorean philosophers: geometry is the key to the Universe.

In every cell of our bodies ROMs are constantly produced by enzyme-catalyzed reactions and are constantly destroyed by enzymes. The ROMs are thus present in a steady state (the rate of formation is equal to the rate of disappearance). Whenever a gene on the DNA, which specifies the synthesis of any of these enzymes is damaged, we create an imbalance between ROM formation and destruction. This imbalance is known as "**oxidative stress**" and leads to pathological consequences.

The complexity of the living cell is enormous. There are about ten thousand different proteins in each cell. Each molecule in a cell has a specific function and must be at the right place, at the right time, so that the correct materials can be manufactured. This is a nightmare for a logistician.

The availability of certain compounds drove evolution in a certain direction, and that's why we are what we are. In the absence of vitamins E and C or lower concentrations of oxygen life would still be possible, but would have led to a different outcome. Our biochemical mechanisms are so finely tuned through billions of years of evolution, that any disturbance will affect our well being. An essential characteristic of living cells is **homeostasis**, which is the maintenance of constant conditions in the internal environment of cells and organs.

We exist within a narrow range of conditions, such as oxygen concentrations, pH (H^+ concentration), calcium ion

(Ca^{2+}), iron ion (Fe^{3+}/Fe^{2+}), potassium ion (K^+), vitamins, and antioxidant concentrations. Everything is poison, it just depends on the dose. Our bodies consist of about 75 trillion cells, organized into different functional units (organs). Each of these structures contributes to the maintenance of homeostatic conditions in the internal environment. As long as normal conditions are maintained all cells and organs will function and prosper. Thus every organ contributes to the well being of the whole, as long as we have cooperation. Every organ has to depend on the other to do its job.

Hemoglobin combines with oxygen as the blood passes through the lungs. The hemoglobin will not release oxygen into the tissue if already too much oxygen is there, but only if the oxygen levels are too low. Hemoglobin thus acts like a buffer to maintain **oxygen homeostasis**.

Carbon dioxide is the end product of combustion engines, but it is also the most abundant end product of our metabolism. Carbon dioxide disolved in water is an acid:

$$CO_2 + H_2O \rightleftarrows H_2CO_3 \rightleftarrows H^+ + HCO_3^-$$

Since our bodies have to maintain a constant pH of about 7.4, the carbon dioxide has to be eliminated. Carbon dioxide excites the respiratory center causing rapid and deep breathing to eliminate the carbon dioxide through the lungs. The maintenance of a constant pH of 7.4 is accomplished via a process known as buffering. This process is based on Chatelier's principle. Blood contains bicarbonate (HCO_3^-, see equation above). Bicarbonate of soda is known to everybody as the ingredient in soda water. The bubbles we observe are carbon dioxide. If we add acid (H^+) to bicarbonate the equilibrium shifts to the left, thus soaking up the excess acid (forming carbonic acid and carbon dioxide). There exist a number of these acid/

base pairs present *in vivo* which buffer our internal fluids and keep a constant pH.

The kidney removes most compounds, which are not needed from the plasma and reabsorbes compounds which can be reused, such as glucose, amino acids, potassium etc.. All the waste products of our metabolism are excreted (in the urine or stool). One of the metabolites of proteins and nucleic acids is ammonia (NH_4OH), which is highly toxic and must be detoxified. This is accomplished by conversion of ammonia to urea and uric acid. Uric acid is an antioxidant. It is present in high concentrations in our plasma. It has been suggested that uric acid has replaced ascorbic acid (Vitamin C) during millions of years of evolution as the most important antioxidants in humans. In converting ammonia to uric acid, we not only eliminate a toxic substance, but at the same time produce something useful (an antioxidant). In the same way society gets rid of its garbage by generating electricity. Although uric acid is an antioxidant, there can be too much of a good thing. Uric acid in excess can lead to kidney and gall stones (gall stones cause abrasion and can lead to cancer).

There are other examples where we encounter too much of a good thing (or too little of a bad thing). One such example is the **SOD paradox**, which I shall discuss in Chapter 4. Excess Vitamin C (too much of a good thing) is catabolized to oxalic acid, which combines with calcium ions (Ca^{2+}) to form the insoluble calcium oxalate (kidney and bladder stones). I shall discuss some more examples of the pros and cons of Vitamin C in Chapter 4.

Excess uric acid is excreted in the urine. Urine consists mainly of water, which is a polar compound, dissolving polar compounds. To eliminate all our metabolic waste products, we have to drink plenty of water. However, even water can kill a cell if it accumulates at the wrong place (edema).

Reproduction.

We have two types of reproduction: asexual and sexual. The simplest living organisms are the single-celled bacteria. Bacteria reproduce by dividing into two cells. Each daughter cell is an exact replica of the original. Unless damaged by outside factors (radiation, heat, chemicals) bacteria can go on dividing forever. They never age nor die.

All of the multicellular organisms reproduce sexually. Sexual reproduction provides some great benefit to the organism: 1. each cell (except sperm and egg cells) contain two copies of each gene. Whenever one gene gets damaged we have another copy. 2. It allows for gene exchange between the two sets. This allows for greater viability in a constantly changing environment.

If the DNA of a bacterium gets damaged, it passes on an exact copy to its offspring. Sexual reproduction allows organism to exchange some genes with its partner and thus alter the genetic make-up. This makes the offspring slightly different from their parents. This is an important benefit in a continually changing environment and is essential for evolution. Otherwise a bad mutation would be passed on and could wipe out the entire species. All multicellular organism have therefore adopted this strategy and reproduce sexually. Sexual reproduction is essential for evolution (survival of the species), but comes at the expense of death for the individual. In biology individuals don't count, they are expendable. Their only function is the reproduction of the genes. Once any organism, be it the fruitfly or a human reaches sexual maturity, it begins to senesce and eventually dies. Our bodies are expendable once they have fulfilled their natural function. All dividing cells in our body eventually stop dividing and die. The number of times a cell divides has become known as the Hayflick limit, after **Leonard Hayflick** who discovered this phenomenon. Each cell has its own Hayflick limit (see Bova, 1998).

Hormones, like estrogen stimulate cell division, and thus provide the conditions for the development of tumors (Henderson et al. 1988). This is why estrogen therapy for post menopausal women is so controversial. Without estrogen women may have hot flashes, experience vaginal dryness, and have a greater risk for osteoporosis, but on the negative side their risk for cancer increases. Are the relative minor side effects of low estrogen worth the risk? I shall come back to this topic in Chapter 6.

We may ask the following question: Does cell division stop because of accumulated damage caused by ROMs or does each cell have a gene in its DNA which tells the cell how often to divide and when to die. We have two types of cell death: **necrosis**, caused by outside agents and ROMs and programmed cell death, which is known as **apoptosis**. The only cells, which do not have a Hayflick limit are cancer cells. Cancer cells are immortal. HeLa cells are cancer cells routinely used all over the world for cancer cell studies. These HeLa cells were taken from the cervical carcinoma of a 30 year old woman named Henrietta Lacks in 1951. When cancer cells grow within our bodies, they keep on growing forever, eventually destroying the host. In cancer cells the apoptotic program is shut off. If we can figure out how to reactivate the apoptotic program, we may be able to stop the cancer.

Economy of design.
Do not make anything yourself, if you can get it for free.

Many chemical building blocks, which we need for the proper functioning of the body, are synthesized by our own metabolism from smaller, simpler molecules. For the synthesis of these many complex biomolecules we need energy (derived from our oxygen metabolism) and we need the software (DNA), which provides our cells with detailed instructions. Just like a computer gets infected with viruses, our software is damaged by reactive oxygen metabolites, the toxic by-products of our power plant (see p. 22).

Any biomolecule, which can be obtained through the food supply and does not have to be synthesized from scratch, saves disk space and energy, and thus confers some advantage to the species using this strategy. Humans cannot synthesize all the 20 amino acids necessary for the built-up of all the proteins and enzymes in our bodies. Eight amino acids have to be obtained through the food we eat and are known as **essential amino acids**. In addition to proteins we need fatty acids for the construction of membranes. There are some fatty acids, which we cannot synthesize, the **essential fatty acids**. These are linoleic acid and alpha-linolenic acid. Since these essential fatty acids are plentiful in our food supply, it would have been foolish to waste valuable disk space and energy, if we can get it for free. This strategy however requires first the evolution of plant and animal species, which serve us as cheap suppliers of important chemicals. The essential fatty acids are the building blocks (parent compound) for a group of other fatty acids. The linoleic acid is the parent of the omega 6- group of fatty acids and the alpha-linolenic acid is the parent of the omega 3-group of fatty acids. These essential fatty acids are abundant in fish, flaxseed, wheat germ and soybean.

Other molecules, which our bodies need but cannot synthesize, are the **vitamins.** Vitamins are essential for many biochemical processes. However these functions are not always related to their antioxidant properties. During the course of evolution, vitamins C and E were plentiful in our food supply and there was no need to synthesize them. Let me point out an analogy between evolution of life and evolution of economics.

Recently economists have argued that what happens in nature also happens in business. The ideas originated largely at the Santa Fe Institute, a cutting edge think tank, which was founded in the mid 1980s (Waldrop, 1992) with the purpose to study complex dynamic systems, such as weather and economics. Corporations shouldn't look to machines as a model, but should look to nature, which provides concrete examples of how to evolve and thrive (Pascale et al. 2000). Recently the Press is full of stories about free trade with China. It makes good economic sense to buy some products cheap from China and incorporate them into high priced, high tech products, rather than make everything ourselves from scratch. A company, which adjusts itself to the China market will prosper, while the others will fail. Remember: anything too inflexible towards change is destined for extinction.

Exchange of genes, in sexually propagating species, is of enormous benefit for the survival of the species and is essential for evolution, where species have to adopt to continually changing conditions. In the same way the exchange of services and ideas improves an economy and civilization.

There are huge numbers of organisms who only reproduce. They do not make anything, they just live off the host continuously reproducing themselves along with a lot of shit. Since they live quite well, they adopt the strategy stated above. A host can very well tolerate a limited number of these individual parasites (it is a question of balance), but everything collapses when their numbers explode.

Another example of the principle of economy of design is the fact, that organisms or organs do not make anything they do not need. Organs exposed to high fluxes of ROMs (such as the lung or liver) have higher levels of antioxidant defenses (SOD, CAT, GSH-Px). This principle, however, does not apply to human society, as the inspection of most peoples closets and garages will demonstrate. You can convince humans (except a few stubborn Swabians) to buy anything they do not need (viruses of the mind).

Molecules, cells, humans and human society.

The most basic things that human beings do are
breath, eat, drink, excrete and have sex.
Paul Davies in "The fifth miracle"

Before getting into scientific details about the chemistry and pathology of ROMs, I like to point out some similarities between the human body and human society. Cells of a human body are organized into organs, which all fulfill special functions. This organization is of great help, not every cell has to do everything. On the other hand this specialization only works if we have cooperation, if every organ in the body can depend on the other to do its job. All cells contain thousands of different molecules, which all have to be at the right place, at the right time and at the right concentrations. We have the same situation in human society. Some people are the brains, some are the police and garbage collectors (phagocytes). Some people pay the taxes (the slaves) and some collect them (the bureaucrats) and some spend them (the politicians). Whenever the politicians spend too much and the people cheat on their taxes we are in trouble. In a complex dynamic system (such as the human body or human society) every section has to do its part for the proper functioning of the whole and any small change can have catastrophic consequences.

Already **Aristotle** thought of the whole universe as one organism. **Kepler** considered the earth as one enormous breathing animal. This idea has recently been resurrected as the **Gaia hypothesis** (Lovelock and Margulis, 1974 and Lovelock, 1979). Damage to any part of the earth can have unforeseen consequences for the well being of the whole. For a number of essays on Gaia, see Margulis and Sagan (1997).

Another interesting analogy between chemistry/biology and human society is the concept of competition. We have competition between species (the driving force of evolution), competition among members of the same species, competition between superorganisms (nations). Competition is also important at the microscopic level between cells (cancer cells versus normal cells, phagocytes versus bacteria). We also have competition between molecules. Antioxidants compete with biomolecules for ROMs. However competition does not stop at the molecular level, but is also present at the submolecular (electronic) level. Although two atoms connected by a chemical bond share the electrons comprising that bond, they do not share them equally. The electrons are attracted more by the more electronegative atom. It is a well established custom among chemists to explain chemical reaction mechanisms by shifting electrons around, as exemplified by the following reactions:

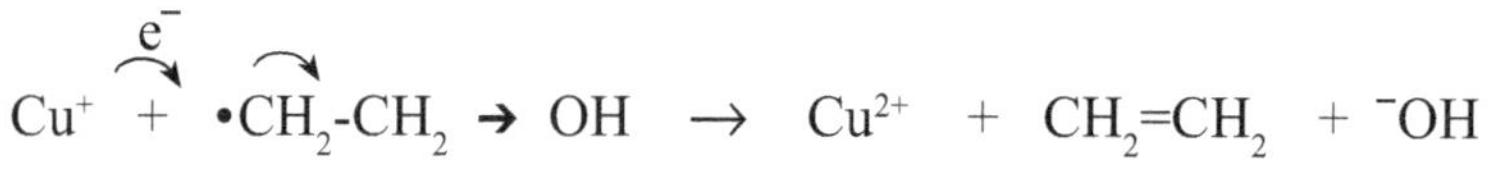

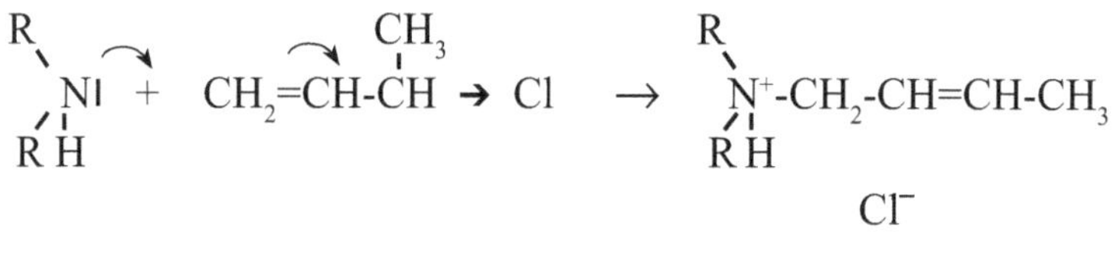

$$Cu^+ \; + \; \bullet CH_2\text{-}CH_2 \to OH \;\; \to \;\; Cu^{2+} \; + \; CH_2\text{=}CH_2 \; + \; {}^-OH$$

$$\begin{array}{c} R \\ \diagdown \\ NI \\ \diagup \\ R\,H \end{array} + \; CH_2\text{=}CH\text{-}\underset{\underset{\displaystyle }{|}}{CH} \to Cl \;\; \to \;\; \begin{array}{c} R \\ \diagdown \\ N^+\text{-}CH_2\text{-}CH\text{=}CH\text{-}CH_3 \\ \diagup \\ R\,H \end{array}$$

$$Cl^-$$

Electrons are the main actors on the stage of life, because electrons are the main actors on the stage of chemistry. A world consisting only of hydrocarbons (-C-C-) would indeed be a dull chemical world. Nothing would change, but nothing would be alive. Change is essential for life.

Protons and electrons are immortal. According to some grand unified theories the proton may be unstable. So far no experimental evidence exists, but some experiments indicate that the life time of a proton is greater than 10^{32} years. Since the age of the universe is about 10^{10} years, we do not have anything to worry about (Lederman, 1993).

Death is a consequence of complexity. Remember **Murphy's Law**: In a complex system, if something can go wrong it will go wrong! In human society, institutions, which do not adapt to changing conditions die (example: communism). The underlying ideas, however, may survive, since they exist as viruses (memes) in our minds. **Martin Luther** saved the Christian Faith from extinction through his reforms. In a democracy we have mechanisms, which adapt the government to changing conditions.

In order for an organism like a human body to function, we need a steady supply of energy (see Brown, 2000). This energy is supplied by the foods we eat and **the oxygen metabolism**. The oxygen metabolism in the mitochondria of our cells charges our battery, which is the adenosine-triphosphate-adenosine-diphosphate (ATP-ADP) couple. ATP is a high energy compound, which is used as a catalyst for many biochemical reactions. We also need the means to supply all our cells with nutrients and oxygen (breathe, eat and drink). These supply lines are the arteries and veins. The arteries deliver the nutrients and oxygen to all our cells and the veins remove the waste products to the garbage dump, the liver and kidneys, where the waste is processed and excreted. The processed waste reenters the biosphere. Nothing ever gets lost, it just changes the molecular structure (reshuffling of electrons). Without liver and kidneys we would drown in our own garbage. We are a throw away society. This is of course a great waste of resources and energy. A living organism cannot afford this luxury. Damaged molecules within a cell are broken down by enzymes to their basic building blocks and reused. *Nature* has discovered recycling long ago. Without recycling of our garbage the superorganism earth (Gaia) will sustain irreparable damage.

In society the supply lines are the highways, airways, waterways and railroads. Whenever any of these supply lines are interrupted (accidents or strikes) we have a cascade of damaging effects. Whenever our oxygen supply lines get obstructed, we need a bypass, otherwise we get a stroke or heart attack. Oxygen is absolutely essential for our survival. All our organs are important for the proper functioning of the whole. Imagine if we wouldn't have truck drivers or airline pilots or garbage collectors or policemen. However, it is the brain, which is the supervisor. The brain, therefore, uses most of the energy we use. Since energy is produced via the oxygen metabolism, it follows that the brain consumes more oxygen than any other organ. The brain consumes about 18% of our total oxygen intake, but constitutes only 2% of total body mass. Interruption of the oxygen supply is the most common cause of death (stroke and heart attack) in the U.S. and other Western countries.

Since the brain consumes more oxygen it also produces more reactive oxygen metabolites ($O_2^{\cdot-}$, $^1O_2$, $\cdot OH$, $HOCl$, $\cdot NO$, ^-OONO). The formation of these reactive oxygen metabolites (ROMs) will be discussed in Chapter 2. In addition, the brain is rich in polyunsaturated fatty acids, which are especially prone towards oxidative damage by ROMs. Any deficiency in antioxidant defenses (superoxide dismutase, catalase, glutathione peroxidase) will affect the proper functioning of the brain. In many neurodegenerative disorders we find an imbalance in these metabolically produced antioxidant defenses. I shall discuss some of these diseases in Chapter 5.

The reactive oxygen metabolites come in two varieties: primary metabolites, which are not very reactive, and secondary metabolites which are highly reactive (Chapter 2). The most reactive oxygen metabolite is the hydroxyl radical or ·OH. The hydroxyl radical reacts with any molecule it encounters. However some molecules are more important than others. We have the same situation in society. It's quite different if somebody kills a policeman or the President of the United States. The most important molecule in our cells is the **deoxyribonucleic acid (DNA).** The DNA is therefore protected by a nuclear membrane and by a protein sheath, the histones. It is therefore impossible for ·OH to penetrate these defenses and attack DNA. The hydroxyl radical does not survive long enough. One of the most important people in the world is the President of the United States of America. The President is well protected by fences, doors and many Secret Service agents. In order to attack DNA, the intruder cannot be a reactive radical like ·OH, but has to be an inocuous small molecule, which is not a radical. This type of molecule is the hydrogen peroxide or H_2O_2. It is stable and easily diffuses through membranes and reaches the DNA. The DNA has an Achilles heel, which is an Fe-counterion. This Fe^{2+} ion can react with the H_2O_2 to form the OH radical in a **Fenton-type reaction,** which I shall discuss in Chapter 2. In this case the hydroxyl radical is formed close to the vital target in what is called a **"site-specific" reaction** (see Eberhardt, 2000). The ·OH, in this situation, cannot be scavenged by added radical scavengers or antioxidants.

The importance of DNA is also evident from the fact, that all our cells contain **two copies of each of our genes**. The importance of a back-up copy is obvious to anybody who ever used a computer and lost a file. Because we have two copies of each gene, these cells are called diploid. One copy comes from our father and one from our mother. Only male sperm and female egg cells are haploid (contain only one copy of each gene).

The observation of a "site-specific" reaction means that we cannot protect DNA from the damage by scavenging the ·OH, but only by removing the H_2O_2. We cannot protect the President of the United States from a bullet fired in close proximity, but we can protect him by eliminating the guns in his proximity (I admit this solution is a little too simplistic).

The Nature of Water and Oxygen

Humankind has been fascinated with water and air throughout the centuries. Among the presocratic philosophers Thales of Miletus believed that the fundamental substance of matter was water, while his fellow philosopher Anaximenes taught that it was air. These philosophers had some deep insight. Obviously water and air are not the fundamental substances of matter, but they are the most important molecules for the evolution of life as we observe it today. Water is a small, low molecular weight molecule, consisting of just two types of atoms, the hydrogen and oxygen atom with the formula H_2O. This is one of the few formulas, which most students remember from their high school chemistry course.

The hydrogen atom is the simplest atom, consisting of one proton (p^+) and one electron (e^-). According to the classical atom theory, we may consider the proton at the center and the electron circling the proton, like the earth is circling the sun. This classical model is known as **the Bohr model**. According to the **quantum-mechanical model**, the electron cannot be pinpointed at a specific position in space (like a planet orbiting the sun), but we can only determine the probability of finding the electron in a certain region of space. This probability distribution is called an orbital. Oxygen has a total of 8 electrons, but only 6 are located in the outer shell, which is involved in chemical bonding. The oxygen atom can only form two bonds involving two of the six outer shell electrons. The other four electrons occupy two orbitals of two electrons each. These pair of electrons have opposite spin and are known as lone pairs (they do not form regular chemical bonds). Since electrons repel each other the hydrogens and the two lone pairs in a molecule of water try to form a three-dimensional structure in which the electrons (orbitals) are as far away from each other as possible. This would lead to a tetrahedral arrangement:

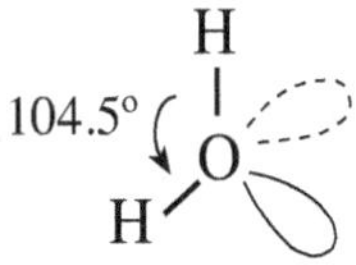

In this arrangement the angle between the two hydrogens and the oxygen should be 109°. In the water molecule this angle is 104.5°. This smaller angle is a consequence of the slightly greater repulsion between the two lone pair electrons. As I have mentioned before a chemical bond consists of pairs of electrons. In a bond connecting two different atoms, the two electrons comprising that bond are shared by the two atoms, but they are not shared equally. The electrons are attracted towards the more electronegative atom:

$$\overset{\delta+}{\text{H}} \rightarrow \overset{\delta-}{\text{OH}}$$

In 1954 **Linus Pauling** received the Nobel Prize in Chemistry for his work on the nature of the chemical bond (Pauling, 1940).

To the general public Pauling is better known for his controversial ideas about ascorbic acid (Vitamin C) and cancer (Pauling, 1970, Cameron et al. 1979). I shall discuss the controversial role of ascorbic acid in more detail in Chapter 4.

The lone pairs of electrons on the oxygen atom of water cannot form regular chemical bonds. The electrical charges, however, attract other water molecules in which the positive H-atom is attracted to the lone pairs. This type of attraction is called a hydrogen bond. The polarity of water makes it an ideal solvent for many polar compounds, such as molecules with C-O, C-N, C-Cl, C-S, or C-P bonds (sugars, amino acids, purines, pyrimidines), and ionic compounds, such as common table salt (Na^+Cl^-). Hydrocarbons such as $CH_3-(CH_2)_n-CH_3$ are non-polar and therefore insoluble in water.

The polarity of the H-O bond is also responsible for the fact that water is a liquid, rather than a gas. Other small molecules, which are non-polar such as O_2, N_2, Cl_2, H_2 are all gases. The water molecule attracts other water molecules via hydrogen bonds and thus forms a well organized macro cluster. In order to transform water into a gas we have to supply energy to break these hydrogen bonds. Because of the nearly tetrahedral arrangement of the electrons around the oxygen atom, each water molecule is capable of hydrogen bonding with four other water molecules.

The cytoplasm of our cells is mostly water, containing thousands of different proteins, sugars, carbohydrates, fatty acids, hormones and nucleic acids. Water is not only a solvent, but is an active ingredient of living systems. Without water surrounding all these molecules, they would loose their biological activity. The folding of a polypeptide into a three-dimensional structure is not guided by a mysterious life force, but by the information encoded in the amino acid sequence when the protein is in water. Hydrogen bonds are also responsible for the formation of double-stranded DNA (see Table III, Appendix). Hydrogen bonds not only form between different amino acids within a long polypeptide chain, but also between $-NH_2$, $-SH$, $-OH$ groups of amino acids and water molecules.

It has been suggested that water on the surface of proteins forms an organized layered structure. Any substance, which dissolves in water, such as a sugar or a NaCl crystal will disturb the extensive dynamic structure of water. The structure of water within the cell is one of the important unsolved problems of biology (see Ball, 1999).

Water in the liquid state is to a small extent dissociated to H^+ (or H_3O^+) and OH^-:

$$H \rightarrow OH \overset{H_2O}{\rightleftarrows} H_3O^+ + OH^-$$

The equilibrium of this equation is far to the left. The pH measures the proton concentration. Since $[H_3O^+]$ is a very small number the pH is defined as the negative logarithm of $[H_3O^+]$:

$$pH = {}^-log\,[H_3O^+]$$

The pH of neutral water is 7.0, i.e. the $[H_3O^+]$ is 10^{-7} moles/liter. In our cells and tissues the pH is 7.4 and may vary only between very narrow limits. Acidosis (pH<7) or alkalosis (pH>8) causes death. A constant pH is maintained with the help of compounds known as buffers.

Another important property of water is the strength of the H-O bond (117.5 Kcal/mole). It is one of the strongest bonds known and can only be broken by high energy radiation. This is, of course, very important to the stability of biological systems. Without this property of H_2O we would not exist.

Another property of water is that it is radioactive (what? and we drink that stuff!). Hydrogen atoms come in three different varieties or isotopes: hydrogen (H), deuterium (D), and tritium (T). The difference between these isotopes is the number of neutrons in the nuclei: 0, 1 or 2 neutrons. This change in the nuclei does not affect its chemistry, since chemistry is controlled by the electrons (reshuffling of electrons). The important point is that tritium is radioactive, and radiation produces OH radicals from water. Fortunately, the amount of tritium in water is extremely low ($10^{-20}\%$), so we don't undergo any major damage. Radicals are the cause of mutations, which in turn are essential for evolution. Without radicals evolution would come to a dead end. Raold Hoffmann in "The Same and not the Same" points out that "pure" water consists of 18 different kinds of molecules (oxygen also exists in three isotopic forms). Hoffmann states, and I quote "The little tritiated water in normal water is something we have evolved with over millions of years. It may even be that the chance variation provided by mutations induced in part by this radioactivity was necessary to get us to the present stage of human creative complexity".

Oxygen is a gas and together with nitrogen (N_2) comprises the major part of the air we breathe (21%). There are two types of oxygen molecules, one which we breath in, and another type, which is highly destructive in living systems. The ordinary oxygen is called the **triplet oxygen ($^3O_2$)** and the highly reactive one is known as **singlet oxygen ($^1O_2$)**. In order to understand the different chemical behavior of triplet and singlet oxygen, we have to understand their electronic structure.

The electron is an elementary particle carrying one unit of negative electrical charge. The electron can be compared to a small sphere spinning around its own axis. The electron can

spin in two directions, clockwise or counter-clockwise. Spinning electrical charges produce a magnetic field, which can be
either up or down ($\uparrow$ or $\downarrow$).

A single C-C bond consists of two electron, spinning in opposite directions. In all neutral stable molecules all the electrons are paired and each pair has opposite spin:

$$\text{-}\overset{|}{\underset{|}{C}}\,\uparrow\downarrow\,\overset{|}{\underset{|}{C}}\text{-}$$

Any fragment, which contains one single unpaired electron is
called a **radical.** Let us look at the electronic structure of an
oxygen molecule (only the electrons in the highest molecular
orbital are shown (a — indicates two electrons):

$$\uparrow\,\overline{\underline{O}}\,-\,\overline{\underline{O}}\,\uparrow$$

Each oxygen carries one unpaired electron with identical spin.
This means that oxygen in its ground state (the stable state) is a
diradical. This state is designated a **triplet state.** Since
radicals are usually very reactive, it is surprising that oxygen
does not react with organic molecules and we do not undergo
spontaneous combustion. Why is this so? In order to form a
chemical bond the two electrons forming this bond must have
opposite spin. This rule follows from the laws of quantum,
mechanics and is known as the **Pauli Exclusion Principle.** To
illustrate this principle I have chosen the reaction of a double
bond with oxygen in a simplified manner:

dioxetane   dioxetane

We can change the spin of one of the electrons via input of energy. The **singlet oxygen** is produced in presence of some dye molecules (such as chlorophyll) and photons (light):

$$D \xrightarrow{h\nu} D^*$$

$$^3O_2 + D^* \rightarrow {}^1O_2 + D$$
$$\text{triplet} \qquad\qquad \text{singlet}$$

where the * indicates an **electronically excited state** of the dye molecule. This process requires photoenergy (hv) and is known as **photosensitization**. Electronically excited states are molecules in which one electron is moved to a higher energy state. This makes the molecule more reactive towards other molecules. The normal way to produce electronically excited states in the laboratory, is by use of light (photons) and is part of the huge field of photochemistry.

The photosensitized formation of singlet oxygen can hardly occur *in vivo*, except in reactions involving epithelial cells. Thus, the discovery of chemical reactions, which produce singlet oxygen is of great importance in biology (examples, see Chapter 2). These reactions give us the possibility of carrying out photochemistry without light. Singlet oxygen reacts with membranes (lipids), proteins and the bases of DNA, and thus damages tissue at various levels.

In the above sequence of reactions we encounter another important group of molecules, the dioxetanes. These dioxetanes have a highly strained four-membered ring, which breaks apart to form two electronically excited states of carbonyl (C=O) compounds:

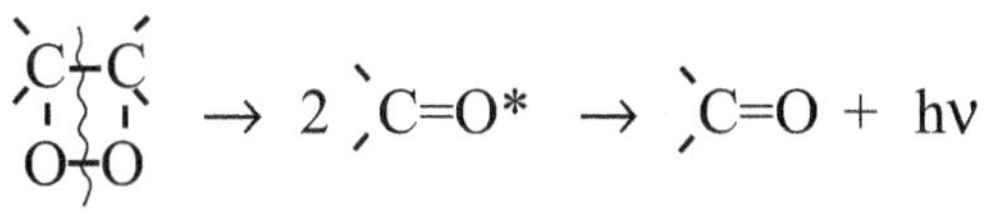

$$\underset{O-O}{C-C} \rightarrow 2\,{\overset{}{\underset{}{}}}C{=}O^* \rightarrow {\overset{}{\underset{}{}}}C{=}O + h\nu$$

The dioxetanes, therefore, provide us with another way to carry out **photochemistry in the dark** (Cilento and Adam, 1988). There are many ways *in vitro* for the formation of dioxetanes, and there is substantial evidence for their formation via enzyme-catalyzed reactions *in vivo*. The above reaction to give dioxetane does not occur with all C=C double bonds. The reaction requires certain substituents on the carbon atoms (see Eberhardt, 2000).

Since oxygen in the ground state is a diradical it can combine with other radicals to form peroxyl radicals:

$$R_2CH{\cdot} + O_2 \rightarrow R_2CH{-}OO{\cdot}$$

These peroxyl radicals are important intermediates in the oxidative degradation of lipids. The ROO radical can undergo dimerization, followed by a break-up to smaller fragments:

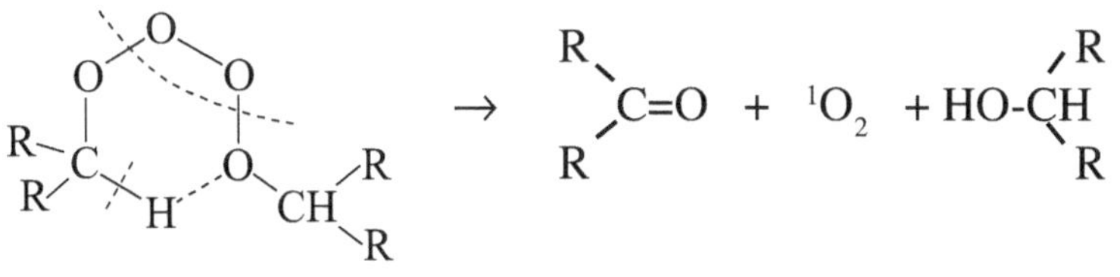

$$\rightarrow \quad \underset{R}{\overset{R}{\diagdown}}C{=}O + {}^1O_2 + HO{-}\underset{R}{\overset{R}{CH}}$$

In this reaction we have converted an unreactive triplet oxygen to the reactive singlet oxygen in the dark. This reaction was first discovered by **Glenn Russell** in 1957 and carries his name. In addition there is evidence for the formation of electronically excited carbonyl compounds (C=O*) in **the Russell reaction**. In damage to lipids these reactions contribute to the damaging effect.

Singlet oxygen is formed via many metabolic processes, especially by activated phagocytes. It can be transformed to the less reactive triplet oxygen by a variety of naturally occurring compounds, called singlet oxygen quenchers. The most efficient singlet oxygen quenchers are the carotenoids (Vitamin A). Carotenoids are also efficient radical scavengers (see Chapter 4).

Chapter 2

REACTIVE OXYGEN METABOLITES

..the main actors of life are electrons.
..... when electrons are passed from donor to acceptor energy
will be liberated and entropy will be increased. It is this that
drives life.
*   Albert Szent-Györgyi in "The living state and cancer"*

Introduction

A chemical bond consists of two or more electrons shared
by two atoms. The electrons forming a stable bond always come
in pairs. A **radical** is a molecular fragment with **one unpaired
electron :**

$$R\text{-}R^{\cdot} \;\rightarrow\; R\cdot \;+\; \cdot R^{\cdot} \qquad (1)$$

Stable molecules are electrically neutral. They contain equal
numbers of positive charges (the nuclei) and negative charges
(the electrons). We can therefore produce a radical either by
removing one electron from a molecule or by adding one
electron to a molecule capable of accepting it. Since water (H_2O)
and oxygen (O_2) are the most important and abundant molecules
of life, let's use these molecules as examples:

$$H_2O \;\rightsquigarrow\; H_2O^{\cdot+} \;+ e^-_{aq} \qquad (2)$$

$$O_2 \;+\; e^-_{aq} \;\rightarrow\; O_2{\cdot}^- \qquad (3)$$

In reaction (2) we obtain a radical with a positive charge
(a cation) and in reaction (3) a radical with a negative charge

(an anion). Reaction (2) takes place under the influence of **ionizing radiation** (represented by the wiggly arrow). As the name indicates ionizing radiation produces ions, in the above case the water radical cation. The electron, which is removed, is surrounded by the polar water molecules, which are attracted to the negative charge. This type of electron is known as a **solvated electron (e^-_{aq})** and was first observed and characterized by Hart and Boag (1962). Since oxygen is an excellent electron acceptor it scavenges the e^-_{aq} produced in reaction (2) and produces a radical anion (the **superoxide radical anion** or frequently called superoxide). The e^-_{aq} can be considered the smallest and simplest radical and demonstrates an important tenet of radical chemistry: a radical reacts with another molecule to give another radical (reaction 3).

The water radical cation ($H_2O^{.+}$) reacts with water to give the hydronium ion and the **hydroxyl radical**:

$$H_2O^{.+} \;+\; H_2O \;\rightarrow\; H_3O^+ \;+\; \cdot OH \qquad (4)$$

This radiation-induced reaction for the formation of hydroxyl radical was established by Lampe and coworkers (1957). In the radiolysis of oxygenated aqueous solutions we produce both $O_2^{.-}$ and $\cdot OH$. However there are procedures available, which allow us to produce exclusively superoxide or hydroxyl radicals (see Eberhardt, 2000).

The superoxide radical anion is the primary reactive oxygen metabolite (ROM). The superoxide radical anion was discovered in 1968 by **McCord and Fridovich** in an enzyme-catalyzed oxidation. The enzyme xanthine oxidase (XO) produces superoxide radical anion:

$$\text{xanthine} \;+\; O_2 \;\xrightarrow{\text{XO}}\; \text{hypoxanthine} \;+\; O_2^{.-} \qquad (5)$$

McCord and Fridovich (1969) also discovered that erythrocytes contained an enzyme, the **superoxide dismutase (SOD)**, which catalyzes the dismutation of superoxide to hydrogen peroxide and oxygen:

$$2 O_2^{\cdot-} \xrightarrow[\text{SOD}]{2 H^+} H_2O_2 + O_2 \qquad (6)$$

This reaction transforms the reactive and damaging super-oxide radical into neutral and less reactive molecules, the hydrogen peroxide (H_2O_2) and oxygen (O_2). It is now known that xanthine oxidase produces both $O_2^{\cdot-}$ and H_2O_2 in equal amounts.

The formation of $O_2^{\cdot-}$ in all aerobic cells and the presence of SOD in most aerobic cells gave rise to the hotly debated hypothesis: $O_2^{\cdot-}$ is toxic and SOD is formed in response to this toxic threat. (Fridovich contra Fee, 1981). The discussion pro and contra the Fridovich hypothesis has at times been rather heated and lasted for many years. In the end the hypothesis espoused by Fridovich has won the debate. I like to limit my discussion to the main question, which was: how super is superoxide? (Sawyer and Valentine, 1981). Chemically (in a test tube) $O_2^{\cdot-}$ is not very reactive. It therefore did not make sense that cells produce SOD to get rid of $O_2^{\cdot-}$, which appeared to be not very reactive. A living organism would hardly waste its resources and energy to produce something it does not need. However we chemists should remember that an aqueous solution is quite different from the internal environment of a cell. I would like to mention just one interesting example, which demonstrates this difference. NADH in aqueous solutions reacts very slowly with $O_2^{\cdot-}$ ($k = 27 \text{ M}^{-1} \text{ s}^{-1}$, Land and Swallow, 1971). However if NADH is attached to the enzyme lac-

48

tate dehydrogenase the reaction proceeds at a reasonable rate (Bielski and Chan, 1973):

$$\text{LDH-NADH} + O_2^{\cdot-} \rightarrow \text{LDH-NAD}\cdot + HO_2^-$$

$$k = 1.0 \times 10^5 \ M^{-1} \ s^{-1}$$

$$\text{LDH-NAD}\cdot + O_2 \rightarrow \text{LDH-NAD}^+ + O_2^{\cdot-}$$
$$k = 10^9 - 10^{10} \ M^{-1} \ s^{-1}$$

We have a chain reaction, leading to high concentrations of H_2O_2.

The ROMs come in several varieties: radicals and non-radical molecular products (such as H_2O_2 and electronically excited states). We also have to distinguish primary and secondary reactive oxygen metabolites as summarized in Fig. 2.

Primary metabolites (low reactivity)
$O_2^{\cdot-}$
H_2O_2
$\cdot NO$

Secondary metabolites (high reactivity)
$\cdot OH$
$^1O_2$
NO_2, N_2O_3, ^-OONO
$ROO\cdot$, $RO\cdot$
$HOCl$

Fig. 2. Reactive oxygen metabolites

I have included in this list of ROMs also **nitric oxide** ($\cdot$NO), which is sometimes designated a nitrogen radical (the unpaired electron spends most of its time at the nitrogen atom). However, since nitric oxide is formed via an oxygen dependent mechanism from the amino acid L-arginine and the oxygen atom in $\cdot$NO is derived from O_2 (Kwon et al., 1990), it may be classified as an oxygen derived radical or reactive oxygen metabolite. In addition the electron in NO is delocalized between the nitrogen and oxygen atoms:

$$\cdot N\text{=}O \quad \longleftrightarrow \quad \ddot{N}\text{-}\dot{O}$$

This delocalization makes the radical more stable (less reactive).

The tools of the chemist

The radiolysis of water is the easiest way to produce the highly reactive hydroxyl radical ($\cdot$OH). Therefore radiation chemists were at the forefront of hydroxyl radical investigations. Radiation chemistry is a relatively new field in the physical sciences. It developed during the Manhattan project of World War II. In 1957 I accepted my first postdoctoral appointment at the University of Chicago to work with Prof. Dr. Weldon Brown on the radiolysis of methanol (CH_3OH). Since as an organic chemist from Germany, I was trained mainly in synthetic procedures, I was first less than enthusiastic to work in radiation chemistry. After irradiation of methanol for hours we obtained mg quantities of ethylene glycol or in presence of oxygen formaldehyde. I asked myself if this procedure would ever be useful for the synthesis of kg. quantities of a useful product. My mind was obviously programmed to think this way (virus of the mind). Nobody could have foreseen the explosive growth of radiation chemistry and its importance in biology and medicine. The radiation chemistry of water laid the groundwork for ROM biology and pathology.

Chemistry consists of synthesis and analysis. Chemists like to play around with solutions, mixing them together to see what is happening. This is the inquisitive kid in all of us. As a teenager I always mixed chemicals and was pleased if the color changed or the mixture got hot or sometimes even exploded. I am sure that most chemists can tell stories about explosions.

Afterwards we have to analyze what kind of new molecule has been formed. Each analysis requires the proper technique. In the old days when I was a graduate student, these techniques were rather simple. The equipment consisted of an analytical balance, a home made melting point apparatus, a fractional distillation setup, a thermometer and your nose. I remember that for my qualitative analysis test, I had to analyze a mixture of five compounds (four liquids and one solid). I still remember three of the liquids, which after fractional distillation (determination of the boiling points) I identified using my nose. Any chemist recognizes the typical smells of cyclohexene, pyridine and ethyl acetate. In order to analyze micrograms or nanograms or less of molecules, and determine these compounds quantitatively, we need equipment more sensitive than a nose. Science aims at a quantitative understanding of phenomena. The development of sensitive analytical techniques is therefore essential for progress in scientific research. Throughout history, scientific progress did not proceed smoothly in a straight line, but in quantum jumps, which occurred after a new technique became available (microscope, electron microscope, X-rays, laser spectroscopy and many more). Some of the techniques which have contributed to the advance of ROM chemistry and biology are: gas chromatography (GC), mass spectrometry (MS), combination of GC-

MS, high performance liquid chromatography (HPLC), HPLC-MS, electron paramagnetic resonance spectroscopy (EPR), spin trapping, microsensors and pulse radiolysis.

These sophisticated analytical instruments make science so expensive. As science pushes the frontiers to extremes, the instrumentation gets more expensive, especially in particle physics. The Fermi accelerator in Batavia, Illinois has a circumference of 4 miles, and the Superconducting Super Collider (SSC), whose construction has been halted for the present, was planned to be of oval shape measuring 53 miles around at an estimated cost in 1992 of more than 8 billion dollars.(S. Weinberg in *Dreams of a final theory*). **Gregor Mendel** did his famous pea breeding experiments, using only his brain and his eyes!

The first radical was discovered by **M. Gomberg** in 1900 (see Ihde, 1967) without today's sophisticated instruments. Gomberg reacted triphenylmethylbromide in benzene solution with silver powder and observed under a carbon dioxide atmosphere that the colorless solution turned yellow, indicating that some reaction had occurred. Upon addition of oxygen, the yellow color disappeared and a peroxide was observed. If the solution was left undisturbed, the yellow color reappeared. From these observations Gomberg postulated a novel species produced via the following reaction:

$$2\Phi_3CBr + 2Ag \rightarrow \Phi_3C\text{-}C\Phi_3 + 2AgBr$$
$$\uparrow\downarrow$$
$$2\Phi_3C\cdot \qquad \text{yellow}$$
$$\downarrow O_2$$
$$\Phi_3C\text{-}OO\text{-}C\Phi_3 \qquad \text{colorless}$$

The discovery by Gomberg was possible because of the stability of the radical. The unpaired electron is delocalized over the three phenyl groups. The delocalized electron gives the triphenylmethyl radical it's yellow color. The above equation demonstrates another radical reaction, namely dimerization: the two unpaired electrons form a new bond. For almost 70 years chemists assumed that the dimer had the simple structure of hexaphenylethane. However in 1968, through the use of nuclear magnetic resonance spectroscopy, the following structure was determined (Lankamp et al., 1968):

$$2\Phi_3C\cdot \rightleftharpoons \Phi_3C \underset{H}{\overset{}{\diagup}} \hspace{-0.5em}\left\langle\begin{array}{c}\\\\\end{array}\right\rangle \hspace{-0.5em}= C\!\!\begin{array}{c}\diagup\Phi\\\diagdown\Phi\end{array}$$

In Gomberg's time any substituent in a molecule was called a radical. Since the triphenylmethyl radical is no longer attached to any other atom or group of atoms, Gomberg called this species a "free radical". This term has unfortunately stuck around until the present time. The adjective "free" should be eliminated. There is no such thing as a "free" radical.

Chemists have arrived at a final theory of chemistry a long time ago. The Schrödinger Wave Equation ($H\psi = E\psi$) describes all of chemistry. In this equation ψ is the wave function of the electrons, H is the Hamiltonian operator (do not worry if you don't know what this means!) and E is the energy level. The equation was first proposed by **Erwin Schrödinger** in 1926, however the solutions to this equation was mathematically so difficult, that it could only be solved for simple species, such as the H_2^+ (hydrogen cation). However, the development of powerful computers and approximations made the solutions possible even for more complex molecules. The field of quantum chemical calculations took off during the last decades. In 1998 the Nobel Prize in Chemistry was awarded to **Walter Kohn and John A. Pople** for quantum chemical theory and the development of computational methods in quantum chemistry. This story again demonstrates the importance of the right tool for the job.

If we study a reaction, we not only want to know what new molecule is formed, but also how it is formed and how fast the reaction proceeds. The former line of inquiry is part of the huge field of reaction mechanisms and the latter is part of reaction kinetics. The speed of a reaction is characterized by its rate constant. These rate constants vary over a wide range. The most reactive radical, the hydroxyl radical, reacts with most molecules with rate constants in the neighborhood of 10^{10} $M^{1}s^{-1}$. Translated into common language this rate constant means that hydroxyl radicals react almost on every collision. Radiation chemists have determined thousands of rate constants for the reaction of hydroxyl radical, using the technique of pulse radiolysis (Buxton et al., 1988). The knowledge of these rate constants is of great importance in determining the effect of any added radical scavenger, which competes with the biomolecule for hydroxyl radicals.

Contrary to the triphenylmethyl radical, most oxygen radicals are very reactive, they have a short life time and are, therefore, difficult to detect. Radicals, however, can be trapped by a spin trap, which converts the radical into another radical, which is more stable and can be measured in an EPR-spectrometer. This methodology has become known as **spin trapping,** as exemplified by the following example:

For a detailed discussion (pro and con) of this and other methods of radical detection see Eberhardt (2000).

At the time of our investigations on the radiolysis of methanol, there was no EPR spectrometer available and spin trapping was not yet invented. In order to prove the intermediate formation of the hydroxymethyl radical, we used nitric oxide ($\cdot$NO) to scavenge the radical and we identified the product:

$$\cdot CH_2OH + \cdot NO \rightarrow ON\text{-}CH_2OH \rightleftarrows HON=CHOH$$
$$\updownarrow$$
$$HONH\text{-}CH=O$$

Unfortunately, the work was never published and we had no idea, that almost 40 years later the radical scavenging reaction of nitric oxide would play such an important role in biology:

$$\cdot NO + O_2^{\cdot -} \rightarrow {}^-OONO$$

The peroxynitrite is the most destructive nitric oxide metabolite. It is one of the destructive agent in activated phagocytes. The hydroxymethyl radical could, of course, have been much easier detected via spin trapping (if it would have been available).

Another interesting example of reaction mechanism is the increased oxygen consumption in activated phagocytes (Sbarra and Karnovsky, 1959). This observation was made long before the superoxide radical anion was discovered in the xanthine oxidase catalyzed oxidation of xanthine (McCord and Fridovich, 1968). The product of this increased oxygen consumption was identified as H_2O_2 (Iyer et al., 1961)

The question now is: how is oxygen transformed into hydrogen peroxide? We can envisage two possibilities (one- or two electron reduction):

$$O_2 + e^- \rightarrow O_2^{\cdot -}$$

$$O_2^{\cdot -} + O_2^{\cdot -} \xrightarrow{2H^+} H_2O_2$$

$$O_2 + 2e^- \xrightarrow{2H^+} H_2O_2$$

As I have mentioned before the $O_2^{\cdot -}$ is a reducing agent. It reduces cytochrome (Fe^{3+}) to cytochrome (Fe^{2+}). The super-oxide dismutase, which catalyses the transformation of $O_2^{\cdot -}$ to H_2O_2 , provides us with a tool to distinguish between the two mechanisms. In presence of SOD, the reduction of cytochrome (Fe^{3+}) will be reduced if $O_2^{\cdot -}$ is an intermediate. This was indeed observed (Babior et al., 1973). In this way the forma-tion of high concentrations of $O_2^{\cdot -}$ was established.

Chemists, especially physical organic chemists, contrary to biologists and biochemists like to work with small molecules. Chemists also work with solutions of organic solvents. These conditions are far removed from biological reality. In 1967 I was offered a position at the University of Puerto Rico Nuclear Center largely because of my previous postdoctoral experience in radiation chemistry. The work carried out in the biomedical building was mainly related to medicine. It dealt with isotope applications to the study of chemical reaction mechanisms, nuclear medicine and radiation therapy. Although I had no direct supervisor and was not told what to do, I decided to make myself useful to the mission of the Institute and I started with irradiating aqueous solutions. Since DNA, a huge molecule containing aromatic bases and sugars, was far too complicated for my taste, I decided to simplify the situation and used benzene (the archetypal aromatic compound), toluene and a number of other mono-substituted benzenes.

Since the radiolysis of water is the best method for the production of hydroxyl radicals, our studies taught us a lot about the properties and reactivity of hydroxyl radicals. We determined the reactivity of hydroxyl towards the different positions (ortho, meta, para) in a monosubstituted benzene. In the hydroxylation of toluene we found that initially almost all the hydroxyl radicals attack the ring positions and do not abstract a H-atom from the methyl side chain. Almost twenty years later the study of OH with nucleosides, nucleotides and DNA (see v. Sonntag, 1987) gave similar results. At least 80% of hydroxyl radicals attack the aromatic bases and only a maximum of 20% attack the sugar.

Although our studies on aromatic hydroxylation were far removed from biological reality, the isomer distribution obtained in the hydroxylation of mono substituted benzenes provided us with a fingerprint for hydroxyl radicals. Almost twenty years later we have used this tool in the identification of hydroxyl radical in the $Cu^+ - O_2$ and in the $Cu^+ - H_2O_2$ reactions (Eberhardt et al., 1989, Eberhardt, 1991). These reactions play a central role in many diseases, such as Alzheimer's, Parkinson's, atherosclerosis and cataract formation.

The primary metabolites, $O_2^{\cdot-}$, H_2O_2 and $\cdot NO$.

The superoxide radical anion is formed in the mitochondria (the power plant) of all cells as a by-product of the energy generating process (the respiratory electron transport chain). ROMs are also formed via many enzymes in the cytoplasm of cells. Some of these enzymes produce both $O_2^{\cdot-}$ and H_2O_2.

The superoxide radical anion is not only a radical, it also carries a negative charge. Opposite charges attract each other, so $O_2^{\cdot-}$ reacts with the ubiquitous H^+ (from dissociated water)

$$O_2^{\cdot-} + H^+ \rightleftarrows HO_2\cdot$$

In this case the reaction can proceed in both directions, we are dealing with an equilibrium. At a pH of 4.8 both $O_2^{\cdot-}$ and $HO_2\cdot$ are present in equal concentrations. Remember: the lower the pH the higher the concentration of H^+. At a pH of normal tissue of 7.4 only about 1% exists as $HO_2\cdot$. The more acid (lower pH), the more we shift the equilibrium to the right. The $HO_2\cdot$ is much more reactive than the $O_2^{\cdot-}$. It abstracts a H-atom from lipids much faster than the $O_2^{\cdot-}$. The $O_2^{\cdot-}/HO_2\cdot$ reacts with lipids (membranes) to produce numerous degradation products, which are toxic and mutagenic.

The primary ROMs are not very destructive themselves, but have to be converted to secondary metabolites. The low reactivity of the primary metabolites is very important, it allows the $O_2^{\cdot-}$ and $\cdot NO$ to also fulfill physiological functions. The $O_2^{\cdot}$ activates genes to produce antioxidants. Nitric oxide ($\cdot NO$) acts as a vasodilator and neurotransmitter. It is the active ingredient in Viagra and increases blood flow to the penis and other organs.

The H_2O_2 because of its small size, electrical neutrality and stability can penetrate cellular walls and therefore act at sites far removed from its site of formation. This is important in the damage caused by hydroxyl radicals to purine and pyrimidine bases (DNA), and I shall come back to this point when I talk about the reactivity of hydroxyl radicals with DNA (pp.77-78).

Another primary oxygen metabolite is the **nitric oxide**. Nitric oxide is a radical (it has one unpaired electron) and this unpaired electron is responsible for its physiological function.

58

Nitric oxide is produced from the amino acid L-arginine and oxygen via an enzyme-catalyzed reaction (Palmer et al., 1988):

$$NOS$$
$$L\text{-arginine} + O_2 \rightarrow L\text{-citrulline} + \cdot NO$$

where NOS is the enzyme **N**itric **O**xide **S**ynthase. Through the use of ^{15}N labeled L-[guanido^{15}N] arginine and gas chromatography-mass spectrometry it was shown that NO is exlusively derived from the terminal guanidine group of L-arginine catalyzed by a number of NO synthases. In addition, it was shown by Kwon et al. (1990) through the use of H_2O^{18} that the oxygen in NO is derived exclusively from molecular oxygen and not from water. The two investigations again show the importance of sophisticated and reliable techniques for the advancement of scientific investigations.

There are many different kinds of nitric oxide synthases produced by different cell types (such as endothelial cells, phagocytes, neurons, microglia). Nitric oxide by itself is not very reactive, but it reacts fast with oxygen to yield nitrogen oxides and with superoxide radical anion to give peroxynitrite ($^-$OONO, see p.63), which is highly destructive. Peroxynitrite reacts with lipids, enzymes and DNA:

see p.63

	lipid	DNA	
aldehydes,	$\leftarrow$ $^-$OONO	$\rightarrow$ 8-OHdG	$\rightarrow$ mutagenesis
ethane, pentane			

$$\downarrow \text{ enzymes containing tyrosine}$$

3-nitrotyrosine

Nitric oxide is the smallest hormone. Due to its small size and electrical neutrality, it easily penetrates cell walls, which is not the case for the negatively charged superoxide radical anion. Nitric oxide fulfills many important physiological functions: Relaxation of vascular smooth muscle, neurotransmission, inhibition of platelet aggregation and phagocytosis. Acetyl-choline, a well known vasodilator, was shown to produce vasodilation *in vivo*, but not *in vitro*. This paradox was resolved by **Furchgott and Zawadzki** (1980). These authors stimulated the aorta with acetylcholine, and observed vasodilation. After the removal of the endothelial cell layer, stimulation no longer produced relaxation. These observations clearly showed the production of an endothelium derived relaxing factor (EDRF) and ended the misconception that the endothelium has no other function but to provide a barrier between cells.

However, nitric oxide also possesses many deleterious effects. It reacts with many biomolecules either by itself or mainly via its oxidation products, the nitrogen oxides (NO_2, N_2O_3, NO_2^-, NO_3^-) or peroxynitrite (^-OONO). Nitric oxide and its oxidation products are part of air pollution and are highly damaging to our lungs. On the other hand, the formation of peroxynitrite (^-OONO) by activated phagocytes is part of the defensive mechanism against invading microorganisms. Macrophages and neutrophils produce high concentrations of both $\cdot NO$ and $O_2^{\cdot-}$, which combine to give peroxynitrite (^-OONO), which is partly responsible for the killing of bacteria and viruses, but which also causes damage to the surrounding healthy tissue. The use of nitric oxide may also save the lives of newborn babies with pulmonary hypertension. Nitric oxide relaxes vascular smooth muscles, allowing more blood to flow This is the effect of Viagra: it increases blood flow to the penis, resulting in an erection. On the other hand, as the body fights infection, phagocytes produce high levels of ·NO, which may lower the blood pressure to dangerously low levels, leading to septic shock. We have again a **pro/con** situation.

Since the discovery of the endothelium derived relaxing factor (EDRF) by Furchgott and Zawadzki in 1980, nitric oxide has been a hot topic. Nitric oxide was chosen in 1992 by *Science* as "Molecule of the Year". Since the discovery of EDRF it took almost a decade before EDRF was definitely identified as nitric oxide.

The vasorelaxing property of ·NO has an interesting history. **Nitroglycerin**, the standard medicine to treat angina pectoris, was already prescribed by nineteenth century physicians. Nitroglycerin is also the active ingredient in dynamite discovered by **Alfred Nobel**. Ironically, Alfred Nobel contracted angina pectoris and was prescribed nitroglycerin, but he refused to take it, being aware of its explosive properties. The mechanism of action of nitroglycerin and ·NO was established by **F. Murad** and coworkers (Arnold et al., 1977, Katsuki et al., 1977) a decade before EDRF was identified as nitric oxide. Murad and coworkers found that nitroglycerin and authentic nitric oxide activated the enzyme guanylate cyclase, which increases the level of guanosine-3', 5'-cyclic mono-phosphate (cGMP). The c-GMP then causes muscle relaxation. A group of Japanese researchers (Miki et al., 1977) made the same observation in cerebral tissue of mice, indicating a role for nitric oxide in the **central nervous system**. As happens often in science, discoveries are made simultaneously and independently by a number of investigators. In 1987 Ignarro and coworkers and Moncada and coworkers (Ignarro et al. 1987, Palmer et al. 1987) published their work, establishing EDRF as nitric oxide. Unfortunately Nobel Prizes can not be shared by more than three people. In 1998 the Nobel Prize in Physiology or Medicine was awarded to **R. Furchgott, F. Murad and L.**

Ignarro for their nitric oxide related work. All one heard in the News was that these scientists discovered Viagra. There is a lot more to nitric oxide than Viagra!

Secondary reactive oxygen metabolites.

Transition metal ions are distinguished from many other ions, by their ability to exist in different oxidation states. These different states contain different numbers of electrons, for example: iron (Fe) can exist as Fe^{3+} and Fe^{2+}, and copper (Cu) can exist as Cu^{2+} or Cu^+. These ions can be converted into each other by the removal or addition of one electron:

$$Fe^{3+} (Cu^{2+}) + e^- \rightleftarrows Fe^{2+}(Cu^+)$$

Fe^{3+} and Cu^{2+} are electron acceptors (they are oxidizing agents) and Fe^{2+} and Cu^+ are electron donating ions (reducing agents). Secondary ROMs are produced from the primary metabolites, and they are highly reactive, causing most of the biological damage, such as attack on lipids, enzymes and DNA. The best known examples of these transformations involve one electron transfer from transition metal ions. These reactions are known as **Fenton-type reactions:**

$Fe^{2+} + H_2O_2 \rightarrow Fe^{3+} + OH^- + \cdot OH$ **Fenton reaction (1894)**

$O_2^{\cdot-} + H_2O_2 \rightarrow O_2 + OH^- + \cdot OH$ **Haber-Willstätter reaction (1931)**

$O_2^{\cdot-} + Fe^{3+} \rightarrow O_2 + Fe^{2+}$ **Fenton reaction driven**
$Fe^{2+} + H_2O_2 \rightarrow Fe^{3+} + OH^- + \cdot OH$ **by $O_2^{\cdot-}$ reduction of Fe^{3+}**

$Cu^+ + H_2O_2 \rightarrow Cu^{2+} + OH^- + \cdot OH$ **Eberhardt et al. 1989.**

The Fenton reaction was first observed by Fenton over 100 years ago (Fenton, 1894). These reactions play a role in Alzheimer's disease, Parkinson's disease, Hodgkin's disease, multiple sclerosis and cataract formation among many others. The Haber-Willstätter reaction is known to most chemists as the Haber-Weiss reaction (1934). However, as pointed out by G. Czapski (1981) this reaction was proposed already three years earlier by Haber and Willstätter (1931). Czapski recommended that the iron-catalyzed Haber-Weiss reaction be called "Fenton reaction driven by $O_2^{\cdot -}$ reduction of Fe^{3+}. There are many examples of misquotes or incorrect nomenclature in the literature. Once a misquote has entered the literature, it is difficult to correct. Most researchers do not always reread the original articles, but rather depend on the statements by others (especially those with a good reputation). I have already mentioned the example of "free radical". It appears that Czapski's recommendation has befallen the same fate. Twenty years after his suggestion the term Haber-Weiss reaction is still in common usage.

The above hydroxyl radical forming reactions are all quite straightforward. The metal ion or the $O_2^{\cdot -}$ donate an electron to H_2O_2, which falls apart to $OH^- + \cdot OH$. However, scientists love to argue about mechanistic details until they are blue in the face, and they want to prove their colleagues wrong. This is why the above simple equations were not accepted right away and numerous other schemes, leading to species different from $\cdot OH$ have been suggested. However,.most scientists today accept the above equations. In order to prove the formation of $\cdot OH$ we need a specific and sensitive analytical tool for the detection of the hydroxyl radical. The pros and cons of hydroxyl radical detection have been discussed in detail by Eberhardt (2000).

A number of other secondary ROMs, which are particularly important in **activated phagocytes and microglia cells** (the phagocytes of the central nervous system) are singlet oxygen ($^1O_2$), hypochlorous acid (HOCl), and peroxynitrite ($^-$OONO). These reactive metabolites are formed via the following reactions:

$$O_2^{\cdot -} + O_2^{\cdot -} \xrightarrow{2\,H^+} H_2O_2 + {}^1O_2$$ dismutation, **Khan, 1970, Khan et al. 1992, 1993**

$$H_2O_2 + Cl^- \xrightarrow{MPO} HOCl + OH^-$$ **Harrison and Schultz, 1976**

MPO: myeloperoxidase enzyme

$$H_2O_2 + HOCl \rightarrow {}^1O_2 + HCl + H_2O$$ **Khan and Kasha, 1963**

$$O_2^{\cdot -} + HOCl \rightarrow \cdot OH + Cl^- + {}^1O_2$$ **Long and Bielski, 1980**

$$\cdot NO + O_2^{\cdot -} \xrightarrow{k} {}^-OONO \xrightarrow{H^+} HOONO \quad pK_a = 6.8$$

$k = 6.8 \times 10^9\ M^{-1}s^{-1}$ **Huie and Padmaja, 1993**

The discovery of singlet oxygen in chemistry and biology has an interesting history. Singlet oxygen was first discovered in the 1920s by astrophysicists. It was largely ignored by chemists until a chance discovery made by Seliger in 1960 (see Kasha, 1979). In the accidental mixing of two solutions (H_2O_2 and HOCl) Seliger observed an orange-red flash in the dark:

$$H_2O_2 + HOCl \rightarrow {}^1O_2 + HCl + H_2O$$

Seliger, however, did not identify the singlet oxygen, but he showed the experiment to Kasha, who just happened to visit his lab. Khan and Kasha then carried out a spectroscopic analysis of the orange-red flash, and identified the singlet oxygen as the intermediate in this reaction (Khan and Kasha, 1963). Before the discovery by Khan and Kasha, singlet oxygen was produced in the laboratory via photosensitization (pp.42,43). This photosensitized reaction is unlikely of any importance within cells, where no light source is available. The discovery of purely chemical reactions for singlet oxygen formation was therefore of fundamental importance to biology. It allows us to carry out **photochemistry in the dark** (Chapter 1). The Khan-Kasha reaction, however is only possible *in vivo*, if we can show that both H_2O_2 and $HOCl$ are formed within cells. The discovery of an enzyme in phagocytes, the myeloperoxidase (MPO), was therefore of great importance. The myeloperoxidase catalyzes the following reaction (Harrison and Schultz, 1976):

$$H_2O_2 + Cl^- \rightarrow HOCl + HO^-$$

Although the reaction of H_2O_2 with $HOCl$ takes place in a test tube, it does not follow that it also occurs *in vivo*. We have to consider many competing reactions:

$$H_2O_2 + Cl^- \xrightarrow{MPO} HOCl + HO^-$$

$$H_2O_2 + HOCl \rightarrow H_2O + HCl + {}^1O_2$$

$$MPO + H_2O_2 \rightarrow \text{deactivation of MPO}$$

The first evidence of singlet oxygen formation in activated polymorphonuclear lymphocytes (PMNs) was presented by Allen et al. (1972), based on chemiluminescence measurements. A few years later Krinsky (1974) showed that bacteria rich in carotenoids (singlet oxygen quenchers) resisted killing by PMNs. Both authors based their observations on the Khan-Kasha reaction. So far, so good. However, Kanofsky et al.(1984), studied the H_2O_2 - Cl^- reaction using purified MPO. The optimum yield of singlet oxygen required a pH of 4 and 5 mM H_2O_2. At higher pH the yield dropped to almost zero. These authors, therefore, concluded that the singlet oxygen forming Khan-Kasha reaction is not an important contributor to the bactericidal activity of PMNs.

In 1989 Kettle and Winterbourne showed that concentrations of 100μM H_2O_2 or higher deactivate MPO. The H_2O_2 concentrations in the phagosome of PMNs is much lower than 100 μM. The absence of singlet oxygen in the previous studies using purified MPO was therefore due to unphysiogially high levels of H_2O_2. This example shows, that one has to be extremely careful in drawing any conclusions about reactions *in vivo* from reactions *in vitro*.

The formation of **singlet oxygen** ($^1O_2$) via dismutation of $O_2^{\cdot-}$ was first postulated by Khan in 1970. This proposal aroused considerable controversy (so what else is new!). It took over twenty years before this reaction was definitely established in activated neutrophils and macrophages by Khan and coworkers (Steinbeck et al., 1992, 1993). I, briefly like to discuss the reasons for this long controversy.

In order to identify the highly reactive singlet oxygen chemists use the reaction of some detector molecule to give a specific and easily detectable product. One such detector molecule is cholesterol, which yields a specific hydroperoxide:

$$\text{Cholesterol} \quad + \quad {}^1O_2 \quad \overset{k_d}{\rightarrow} \quad \text{ROOH}$$

In order for this reaction to proceed, it has to be fast enough to compete with another important reaction of singlet oxygen. This reaction is called quenching and transforms singlet oxygen to the unreactive triplet oxygen:

$$ {}^1O_2 \quad + \quad Q \quad \overset{k_q}{\rightarrow} \quad {}^3O_2 \quad + \quad Q^* $$

Singlet oxygen reacts with PUFAs, enzymes and DNA leading to hydroperoxides, endoperoxides, dioxetanes, aldehydes and hydroxylated DNA bases (8-OHdG). Singlet oxygen quenchers are therefore important antioxidants (see chapter 4 under carotenes).

The dismutation of $O_2^{\cdot-}$ requires high concentrations of superoxide in order to compete with all the other reactions of $O_2^{\cdot-}/HO_2\cdot$ (such as lipid peroxidation). Superoxide is also an efficient singlet oxygen quencher, and the product forming reaction has to compete with the quenching reaction. Since in activated phagocytes the concentration of $O_2^{\cdot-}$ is high and the rate constant of $^1O_2$ with cholesterol is low, the conditions for singlet oxygen detection are unfavorable. Early experiments using cholesterol as a singlet oxygen probe, therefore gave negative results (Foote et al., 1981). In later experiments Khan and coworkers (Steinbeck et al., 1992, 1993) used 9, 10-diphenylanthracene as a singlet oxygen probe. This reactions gives 9,10-diphenylanthracene endoperoxide and is considerably faster than the reaction with cholesterol. Khan and coworkers identified the DPA endoperoxides. In presence of ß-carotene (an efficient singlet oxygen quencher) no DPA endoperoxide was formed, thus clearly establishing singlet oxygen in activated neutrophils and macrophages.

Another singlet oxygen forming reaction is the reaction discovered by Long and Bielski (see p. 63). This reaction is the only OH radical forming reaction, which does not require metal catalysts. It is of importance in activated phagocytes.

The **hydroxyl radical ($\cdot$OH)** is the most destructive radical known. It reacts with proteins (enzymes), lipids (membranes) and DNA. In addition to the above reactions the hydroxyl radical is formed via the **radiolysis of water.** An infamous example of the destructive power of the OH radical is Hiroshima and Nagasaki. Hydroxyl radicals however can also be put to some good use, such as in the **radiation therapy of cancer.** On the other hand, radiation causes cancer. Hydroxyl radicals are also important in the killing of bacteria by phagocytes (Chapter 3). We encounter again a **pro and con** situation.

Iron ions (Fe^{2+}) and copper ions (Cu^+) are responsible for the formation of the most damaging radical, the hydroxyl radical. These metal ions are however very essential for our well-being. The iron is needed for **hemoglobin**, the molecule which transports oxygen to every part of our bodies. Copper, iron, manganese, zinc are needed for the synthesis of the antioxidant enzyme, the **superoxide dismutases (SODs)**. So, iron or copper ions can be good or bad, depending on the specific circumstances. The Fe^{2+}/Cu^+ ions are not only involved in the formation of the hydroxyl radical, but also in the deactivation of $O_2^{\cdot-}$ via the SOD enzymes. Iron and copper ions are part of the active sites of the SOD enzymes. These enzymes shuttle electrons between $O_2^{\cdot-}$ and the metal ions (Klug et al., 1972):

$$Cu^{2+}ZnSOD \; + \; O_2^{\cdot-} \; \rightarrow \; Cu^+ZnSOD \; + \; O_2$$

$$Cu^+ZnSOD \quad + \quad O_2^{\cdot-} \quad \xrightarrow{2H^+} \quad Cu^{2+}ZnSOD \quad + \quad H_2O_2$$

Due to the importance of some **transition-metal ions** in hydroxyl radical formation it is not surprising to find a relationship between iron body stores and **colorectal and lung cancer** (Stevens et al. 1988, Knekt et al. 1994). On the basis of these results one may be tempted to assume that high body iron stores contribute to other diseases as well. Your mother's admonition, popularized by the cartoon character Popeye, to eat your spinach, may not be such good advice. However, like everything else in life, it is a question of balance.

The importance of iron in oxidative injury has been reviewed by McCord (1991, 1998). There is considerable evidence showing that **iron overload** is a risk factor in many diseases, such as colorectal and lung cancer, heart disease (Salonen et al. 1992) and Parkinsons disease. It has been proposed by Sullivan (1981, 1989) that even normal iron sufficiency imposes a risk factor and explains the gender difference in death rates from ischemic heart disease. Healthy young men have serum ferritin levels three or four times those of healthy young women. It has been suggested that menstrual blood loss (resulting in lower serum and tissue ferritin stores) rather than high estrogen levels, may be the protective factor.

Blood donation three times a year will lower the serum ferritin level of a man to that of a young woman. Medieval physicians prescribed blood letting for almost anything which ailed you. Maybe there is some scientific basis for this procedure, which may turn out not to be as foolish as it first appears. Regular blood letting in men may lower the risk for colorectal and lung cancer, heart disease and Parkinson's disease. The Red Cross slogan:"donate blood and give the gift of life" should be addressed to males: "donate blood and the life you save may be your own". In addition to increased risk of certain cancers, in an epidemiological study on a large number of Finnish men (Salonen et al. 1992), it was determined that high levels of iron in serum correlated with increased risk of coronary artery disease.

The most important parts of the cell are the DNA and the membrane, which protects the integrity of the cell. Some of the ROMs react with these molecules with damaging consequences. However, the attack of radicals on DNA is essential for mutation and evolution. Without radicals evolution would come to a dead end. We again have a **pro/con** situation.

As already mentioned ROMs show different reactivity. Some of them (**·OH, $^1O_2$, $^-$OONO**) react very rapidly with lipids (membranes), proteins (enzymes) and most important with DNA, causing serious damage. The hydroxyl radical reacts with any molecule it encounters, mostly via abstraction of a H-atom or via addition to an electron-rich molecule, like aromatic compounds. The four bases of DNA are such aromatic compounds.

The ROMs are produced in many different kind of cells. ROMs are produced in mitochondria (the energy plant of the body) of all cells, in epithelial cells (· NO) and neurons (· NO). However the most prolific producer of ROMs are the phagocytes (Chapter 3). The phagocytes are the police and the garbage collectors of the body. They defend us against bacteria and viruses and are absolutely essential for our survival. Patients with defective phagocytes (chronic granulomatous disease) die in infancy. I shall discuss the **pros and cons** of phagocytes in Chapter 3.

It is interesting to note, that all ROMs are small, simple radicals or molecules, which are present in many common products. Hypochlorous acid (HOCl) is the active ingredient in bleach and hydrogen peroxide (H_2O_2) is added to toothpaste, because of its antibacterial effect. The HOCl is also used to disinfect swimming pools. Nitric oxide is the latest hit of the pharmaceutical industry (Viagra).

Reactions of oxygen derived radicals.

Important biological objects come in pairs (two eyes, two lungs, two kidneys, two halves of our brain, two strands of DNA, two sexes). In chemistry **two electrons** form a stable single bond.

Since radicals have one unpaired electron, they are very reactive and try to find another electron with which they can pair. A radical can either obtain another electron or donate its unpaired electron to another molecule or radical, in other words a radical can act as an oxidizing or a reducing agent:

$$O_2 \xleftarrow{-e^-} O_2^{\cdot -} \xrightarrow[\cdot 2H^+]{\cdot e^-} H_2O_2$$

Some examples are:

$$O_2^{\cdot -} + Fe^{3+} \rightarrow O_2 + Fe^{2+}$$

$$O_2^{\cdot -} + H_2O_2 \rightarrow O_2 + OH^- + \cdot OH \quad \textbf{Haber-Willstätter}$$
$$\textbf{reaction (1931)}$$

$$O_2^{\cdot -} + Cu^+ + 2\,H^+ \rightarrow H_2O_2 + Cu^{2+}$$

A reaction in which a radical acts as both oxidizing and reducing agent is called **dismutation**:

$$2\ H^+$$
$$O_2^{\cdot -} + O_2^{\cdot -} \rightarrow H_2O_2 + O_2$$

In this reaction one $O_2^{\cdot -}$ acts as an electron donor to give O_2 and the other acts as an oxidizing agent (an electron acceptor) to give H_2O_2. The hydroxyl radical ($\cdot OH$), which is much more reactive than the superoxide radical anion ($O_2^{\cdot -}$) is also highly electrophilic (Eberhardt, 1977), i.e. it reacts with other molecules as an oxidizing agent (it accepts an electron or H atom). This can be accomplished either via abstraction of a hydrogen (H) atom or via addition to electron-rich aromatic compounds, such as the bases of DNA (using guanine as example):

$$\cdot OH + AH_2 \rightarrow H_2O + AH\cdot$$

$$\cdot OH + \alpha\text{-TOH} \rightarrow H_2O + \alpha\text{-TO}\cdot$$

$$\cdot OH + LH \rightarrow H_2O + L\cdot$$

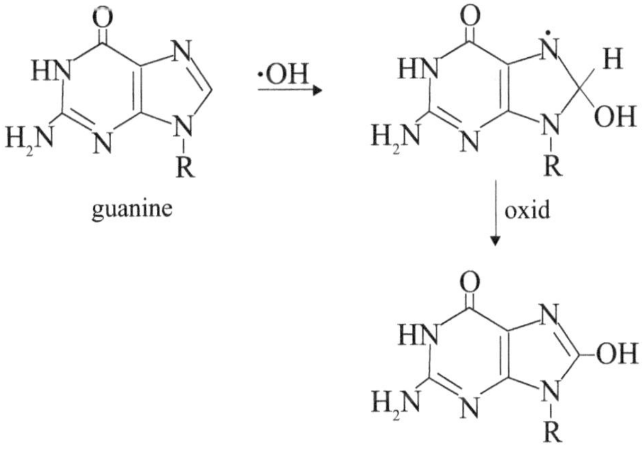

guanine

oxid

8-OHd G

where AH_2, α-TOH, LH represent ascorbic acid (Vitamin C), α-tocopherol (Vitamin E) and lipid (LH) respectively. These reactions of hydroxyl radicals (·OH) demonstrate an important tenet of radical chemistry: a radical reacting with a molecule always begets another radical. In the above examples the highly reactive hydroxyl radical is converted to the less reactive radicals AH·, α-TO· and L·. Ascorbic acid (Vitamin C) and α-tocopherol (Vitamin E) offer protection against radical in-duced damage through **self-sacrifice.** This is why patients, who have some disease caused by oxidative stress, have low levels of antioxidants in their plasma. Vitamins which are water soluble, such as Vitamin C have to be taken daily, since they are consumed by ROMs and any excess is excreted. Lipid soluble vitamins, such as vitamins A or E are stored in the fat and provide us with a longer lasting supply.

The addition of ·OH to aromatic molecules was first demons-trated in the radiolysis of aqueous benzene solutions, using the technique of pulse radiolysis (Dorfman et al., 1962):

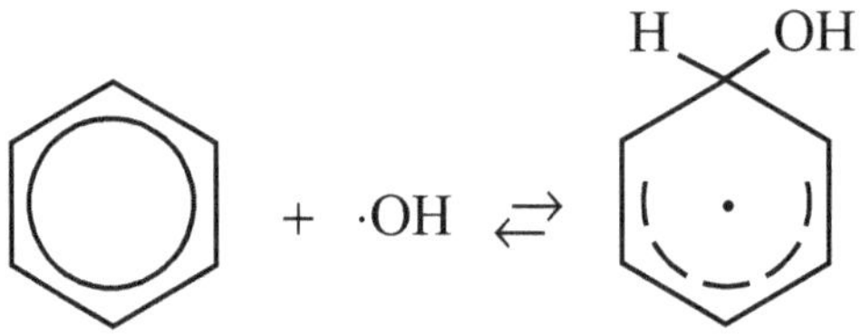

Simultaneously the same intermediate was observed in the Ti^{3+} -H_2O_2 + benzene reaction using electron paramagnetic resonance spectroscopy (Dixon and Norman, 1962). These experiments showed that in the Ti^{3+} -H_2O_2 reaction the ·OH was the reactive intermediate. Almost twenty years later, it was shown that the above addition reaction is reversible (Eberhardt, 1981).

Another important reaction of some radicals is the combination with another radical (O_2 is a diradical!):

$$R \cdot + O_2 \rightarrow ROO \cdot$$

This reaction is important in the chain oxidation of unsaturated fatty acids and was already discovered by Gomberg. Another example is the combination of $\cdot NO$ with $O_2^{\cdot -}$ to give the highly damaging peroxynitrite anion (p. 65):

$$O_2^{\cdot -} + \cdot NO \rightarrow {}^- OONO$$

In order for this reaction to occur it has to compete with other reactions of $O_2^{\cdot -}$ and $\cdot NO$, such as initiation of lipid peroxidation and oxidation of $\cdot NO$ to nitrogen oxides (NO_2, N_2O_3). Therefore, both $O_2^{\cdot -}$ and $\cdot NO$ must be formed at the same location and in high concentrations. These conditions are fulfilled in activated phagocytes (Chapter 3).

The three principal mechanisms for cell injury and tissue damage induced by radicals and electronically excited states are: damage to hardware (lipids, proteins), software (DNA/RNA) or to the power plant (mitochondria). The damage can be caused directly by the ROMs through the direct attack on these essential targets. The damage may be mediated by the destruction of the lipid bilayer via **lipid peroxidation** or the damaging effect may be due to secondary reactions of the lipid degradation products (aldehydes) with critical targets (enzymes, DNA). DNA contains the instructions for protein synthesis. Any damage to DNA means a misspelling in the message and thus a damaged protein and a metabolic defect, unless the damage can be repaired. The enzyme DNA polymerase constantly patrols along the DNA chain looking for damaged bases, which are then excised and repaired.

Since ROMs are formed via normal metabolism (oxidative phosphorylation, phagocytosis, neurons, endothelial cells), we observe tissue damage even in the absence of outside toxic influences. It has been estimated that ROMs damage our DNA about 10^4 times/day. I am now going to discuss some of these ROM-induced damaging reactions.

The membrane of cells and subcellular organelles is of utmost importance to the viability of cells and their organelles. The most significant function of membranes is to provide a barrier. The membrane keeps certain compounds in and others out. Without the membrane the many reactions taking place within a cell would lead to chemical chaos and eventual cell death. In order to fulfill their physiological functions the membranes have to be flexible. This flexibility is accomplished through unsaturation.

Unfortunately the unsaturation, which makes the membrane flexible also makes it more prone towards oxidative damage. Food chemist have been interested in the oxidation of fatty acids, since this process leads to rancification through degradation. The important PUFAs all have one or more $-CH_2-$ groups adjacent to two olefinic bonds, i. e. the structural unit $\sim CH=CH-CH_2-CH=CH\sim$. Both the hydroxyl and the perhydroxyl radical can abstract a hydrogen from these methylene ($-CH_2-$) groups. These reactions are summarized as follows:

$$LH + \cdot OH \ (or \ HO_2 \cdot) \ \rightarrow \ L \cdot + H_2O \ (or \ H_2O_2) \quad \text{initiation}$$

$$L \cdot + O_2 \ \rightarrow \ LOO \cdot$$

propagation

$$LOO \cdot + LH \ \rightarrow \ LOOH + L \cdot$$

$$LOO \cdot + \cdot OOL \quad \rightarrow \quad LOOOOL \quad \text{termination}$$

A lipid radical L· reacts with oxygen to give a peroxyl radical (LOO·), which in turn produces another lipid radical L·, in other words we have a **chain oxidation**. The LOOH is however not the final stable product.

The lipid hydroperoxides subsequently decompose to give a huge variety of aldehydes:

$$LOOH + Fe^{2+} \quad \rightarrow \quad LO \cdot \quad + OH^- \quad + \quad Fe^{3+}$$

$$R\text{-}CH\text{-}CH\text{=}CH\text{-}CH\text{=}CH\text{-}R^{\cdot} \rightarrow R \cdot + OCH\text{-}CH\text{=}CH\text{-}CH\text{=}CH\text{-}R^{\cdot}$$
$$|$$
$$O \cdot \quad \downarrow$$
$$R\text{-}CHO + \cdot CH\text{=}CH\text{-}CH\text{=}CH\text{-}R^{\cdot}$$

LOOH reacts with metal ions (Fe^{2+}, Cu^+) in a Fenton type reaction (electron transfer) to give alkoxyl (LO·) radical. The initially formed radical LO· falls apart to give a stable molecule (an aldehyde) and another, smaller radical. Aldehydes (R-CH=O) react with many biomolecules and cause serious cellular damage. Another reaction forming alkoxyl radicals is the reaction of lipid hydroperoxides with Vitamin C (Lee et al., 2001). This reaction appears not to involve transition metal ions and will certainly add additional controversy to the Vitamin C story. I shall come back to the pros and cons of Vitamin C in Chapter 4.

These aldehydes are mainly responsible for the toxic effect of lipid peroxidation. Unsaturated aldehydes can undergo Michael addition:

$$RCH(OH)\text{-}CH\text{=}CH\text{-}CHO + H\text{-}X \rightarrow RCH(OH)\text{-}CH(X)\text{-}CH_2\text{-}CHO$$

or they can form Schiff bases:

$$RCHO + R'NH_2 \rightarrow RCH=NR' + H_2O$$

The Schiff bases in turn can undergo oxidation by peroxidase enzymes to give dioxetanes (Chapter 1), which are known to decompose to give electronically excited carbonyl compounds.

$$RCH_2CH=NR' \rightleftharpoons RCH=CH-NHR'$$

$$O_2 \downarrow peroxidase$$

$$\underset{O-O}{RCH-CH-NHR'} \rightarrow RCHO^* + R'NHCHO$$
$$\qquad\qquad\qquad\qquad triplet$$

These electronically excited states can carry out **photochemistry in the dark.** This of course is of tremendous importance within cells where no light is available. These excited states damage proteins and DNA. In addition the aldehydes are chemotactic, they attract more phagocytes to the affected site, thus aggravating the damage. We have the following sequence of events:

activated neutrophils → inflammation → lipid peroxidation →

release of hydroxyalkenals → attraction of more neutrophils

Aldehydes are the major products of lipid peroxidation. The most biologically important aldehydes are malonaldehyde and 4-hydroxy-2-nonenal. Aldehydes are involved in mutagenesis and tissue damage., and their chemical and biological properties have been extensively investigated. In addition to aldehydes, lipid peroxidation has been shown to yield ethane as well as other alkanes and alkenes. The determination of ethane (CH_3CH_3) and pentane (C_5H_{12}) in exhaled breath has been suggested as a measure of lipid peroxidation *in vivo* (Riely et al. 1974, Muller and Sies, 1984). I shall come back to these markers again in Chapter 5 under exercise.

Science makes everything easier to understand. Life is not complex, it just appears that way. The underlying laws are simple. Sometimes we get lost in the immense sea of information and get too engrossed in details, and lose sight of the whole. The book of life is written in just four letters, the DNA bases adenine (A), guanine (G), cytosine (C) and thymine (T). Combinations of three of these letters specify all the 20 amino acids, necessary for the synthesis of all the proteins in our bodies. Any damage to the four bases will cause misspelling of the instruction and thus lead to pathological changes.

The species responsible for DNA damage are not $O_2 \cdot^-$ and H_2O_2 (the primary metabolites), but the hydroxyl radical, singlet oxygen, electronically excited states, and peroxynitrite produced via secondary reactions. The DNA is well protected by the nuclear membrane and by the surrounding proteins (histones). The hydrogen peroxide is the only neutral metabolite which can migrate from its site of formation and react at distant sites. It penetrates the membrane and interacts with DNA. In order to cause damage, it reacts with Fe(II), which is present as a counterion in DNA. The resulting hydroxyl radical reacts close to the site of formation in a site-specific manner The same **site-specific damage** has been observed in proteins (Stadtmann and Oliver, 1991).

Hydroxyl radicals can react with DNA in two ways: 1. Addition to the purines or pyrimidines, and 2. Hydrogen abstraction from the sugar-phosphate backbone. Since ·OH is a highly electrophilic radical (Eberhardt, 1977), it will preferentially attack the electron-rich rings of the pyrimidines and purines. The amount of attack on the sugar-phosphate backbone does not exceed 20% (v. Sonntag, 1987, Dizdaroglu, 1993).

The oxidative damage to DNA bases has been considered to be a significant source of **mutations** and many degenerative diseases like aging and cancer (Ames and Shigenaga, 1993). The damaged DNA is constantly repaired and the damaged bases are excreted in the urine. One such oxidation product is **8-hydroxydeoxyguanine (8-OHdG).** The biologically most important property of 8-OHdG is its mutagenicity (Cheng et al. 1992).

The formation of 8-hydroxydeoxyguanosine was observed in the test tube via many different chemical cocktails, which produce hydroxyl radicals. It is produced by j-irradiation (as in radiation therapy) and X-irradiation (be aware of too many X-rays over a short period of time). Although the hydroxyl radical (·OH) gives 8-OHdG, the formation of 8-OHdG is no proof of ·OH. Singlet oxygen and electronically excited states and peroxynitrite ($^-$OONO) also react with DNA to give high yields of 8-OHdG (see Eberhardt, 2000):

$$^1O_2$$
$$·OH$$
$$DNA \quad \rightarrow \quad 8\text{-OHdG}$$
$$^-OONO$$

The 8-OHdG is by no means the only degradation product of DNA. However, due to the availabilty of an easy and sensitive analytical technique (Floyd et al. 1986) for the detection of 8-OHdG this product has become the standard marker for oxidative damage.

The peroxynitrite anion plays a central role in oxidative damage. This compound is formed by the rapid combination of

·NO and $O_2^{·-}$ (see p. 63) and is especially formed in high concentrations by activated phagocytes.

The mechanism by which peroxynitrite exerts its damaging reaction has been extensively investigated and debated. It has been suggested that peroxynitrite decomposes to give hydroxyl radicals:

$$^-OONO \; + \; H^+ \; \rightarrow \; HOONO \; \rightarrow \; ·OH + ·NO_2$$

At present the formation of hydroxyl radicals appears unlikely. A detailed discussion of this problem can be found in the book by Eberhardt (2000).

Peroxynitrite reacts with the amino acid tyrosine and tyrosine containing proteins to produce **3-nitrotyrosine**. The 3-nitrotyrosine is a stable product, which is excreted in the urine. It has, therefore, been suggested as a marker for total oxidative damage caused by reactive nitrogen species (RNSs). 3-Nitrotyrosine was found in blood serum and in synovial fluid of patients with **inflammatory joint disease** (rheumatoid arthritis). Blood and synovial fluid from normal subjects contained no detectable levels of 3-nitrotyrosine (Kaur and Halliwell, 1994). The 3-nitrotyrosine can also be formed by attack of hydroxyl radicals on tyrosine containing proteins. The observation of the 3-nitrotyrosine in serum or in urine is therefore indicative of oxidative stress, but does not tell us what mechanism is responsible.

Another important group of biomolecules are -SH containing proteins and enzymes, such as **glutathione peroxidase**, which eliminates H_2O_2 at the expense of GSH and formation of GSSG. The GSSG is reduced back to GSH by the **glutathione reductase**, which uses NADPH as the hydrogen donor:

$$\text{H}_2\text{O}_2 \ + \ 2\,\text{GSH} \ \overset{\text{GSH-Px}}{\rightarrow} \ 2\,\text{H}_2\text{O} \ + \ \text{GSSG}$$

$$\text{GSSG} \ + \ 2\,\text{NADPH} \ \overset{\text{Glu-red.}}{\rightarrow} \ 2\,\text{GSH} \ + 2\,\text{NADP}$$

Therefore, the formation of GSSG has been used as a marker for increased production of reactive oxygen metabolites. The GSH/GSSG ratio is normally high to protect cells from oxidative stress. In liver, the ratio has been found to be ca. 300/1. Defenses are higher in organs exposed to high fluxes of ROMs (economy of design).

The glutathione reductase produces GSH (an antioxidant) from GSSG and thus offers protection against oxidative damage. The increased formation of $O_2\cdot^-$ (a prooxidant) stimulates the formation of Glu-red and thus increases the level of GSH (an antioxidant). The cells are not defenseless against the increased formation of $O_2\cdot^-$, but respond with the synthesis of antioxidants. I shall come back to this problem later on under **SOD paradox** and under **exercise**.

Summary.

In order to determine if our cells and tissues undergo oxidative damage, we need a specific, sensitive and reliable analytical tool. The damage to lipids is determined by aldehydes, ethane and pentane. The damage to DNA is determined by **8-OH-dG** and the damage to proteins containing tyrosine is determined by 3-nitrotyrosine. The determination of the GSH/GSSG ratio is a measure of total oxidative damage. An additional indicator of oxidative damage is the determination of allantoin, which is an oxidation product of uric acid. Uric acid is the most abundant antioxidant in human plasma and will be discussed in Chapter 4.

Chapter 3

THE PROS AND CONS OF PHAGOCYTES

Introduction

In human biology the functions of police and garbage collector are carried out by leukocytes or white blood cells. The leukocytes are the body's quick response team against foreign invaders. They are formed partially in the bone marrow and partially in the lymph tissue, and they are transported in the blood to different parts of the body, where they are needed. The white blood cells are not just randomly circulating in the blood, but are specifically transported to areas of inflammation, thus providing a quick response to any invading bacteria, viruses or toxin. How do the leukocytes know where to go? Phagocytes are attracted to the point of invasion partly by electrical forces. In addition phagocytes are attracted by chemical substances. This latter process is known as chemotaxis. In mechanical trauma (crushing of the tissue), the cells spill their content into the extracellular environment, releasing iron ions from their storage proteins, initiating oxidation of polyunsaturated fatty acids (PUFAs). These oxidations produce a great variety of aldehydes (see Chapter 2), which are chemotactic. We have the following sequence of events:

Crushing of tissue → release of iron ions from their storage proteins → oxidation of lipids to aldehydes → attraction of more phagocytes → formation of more ROMs → oxidation of more lipids to aldehydes.

There are several different types of leukocytes, distinguished by their different morphology. The most abundant leukocytes are the **neutrophils** and the **macrophages**, which digest the intruder via a process called **phagocytosis**. The neutrophils, produced in the bone marrow, are already mature cells and can immediately begin phagocytosis. The monocytes on the other hand, are still immature cells when discharged from the bone marrow. They slowly mature and become macrophages. The first line of defense against infection, are therefore the neutrophils, while the macrophages are a more powerful second line of defense. Phagocytes (neutrophils and macrophages) continue to digest bacteria, foreign particles, cellular debris and chemical toxins until they die themselves. A neutrophil usually is capable of destroying 5-25 bacteria before being killed themselves. The macrophages are much more effective, killing up to 100 bacteria before death. The leukocytes make the ultimate sacrifice for the well being of the whole organism.

As a neutrophil approaches a bacterium it projects pseudopodia in all directions (like an octopus). These pseudopodia surround the bacterium completely, thus forming an enclosed space. The neutrophils and macrophages contain many lysosomes, which release their enzymes, which digest proteins and cell walls (lipids). In addition to these lysosomal enzymes phagocytes produce ROMs in a process known as the respiratory burst. The contact between a bacterium and a phagocyte stimulates enzymes (the NADPH oxidase) within the phagocyte's membrane. This causes increased oxygen consumption and production of $O_2^{\cdot-}$ and H_2O_2:

$$\text{NADPH} + 2\,O_2 \quad \xrightarrow{\text{NADPH oxidase}} \quad \text{NADP}^+ + 2\,O_2^{\cdot-} + H^+ \quad (1)$$

Another important enzyme in phagocytes is the **myelo-peroxidase** (MPO), which catalyzes the following reaction (Harrison and Schultz, 1976):

$$\text{MPO}$$
$$H_2O_2 + Cl^- \rightarrow HOCl + OH^- \tag{2}$$

Stimulated phagocytes also produce increased levels of nitric oxide ($\cdot NO$), catalyzed by some **Nitric Oxide Synthase** (NOS):

$$\text{NOS}$$
$$\text{L-arginine} + O_2 \rightarrow \text{L-citrulline} + \cdot NO \tag{3}$$

However, as I have already pointed out these primary metabolites are not very reactive. They have to be converted first to more reactive secondary ROMs :

$$O_2^{\cdot-} + O_2^{\cdot-} \xrightarrow{2\,H^+} \mathbf{H_2O_2} + {}^1\mathbf{O_2} \qquad \text{dismutation} \tag{4}$$

$$H_2O_2 + Fe^{2+} \rightarrow Fe^{3+} + OH^- + \cdot\mathbf{OH} \quad \text{Fenton reaction} \tag{5}$$

$$O_2^{\cdot-} + Fe^{3+} \rightarrow O_2 + Fe^{2+} \qquad\qquad \text{Fenton reaction driven} \tag{6}$$
$$Fe^{2+} + H_2O_2 \rightarrow Fe^{3+} + OH^- + \cdot OH \quad \text{by } O_2^{\cdot-} \text{ reduction of } Fe^{3+}$$

$$H_2O_2 + HOCl \rightarrow {}^1\mathbf{O_2} + HCl + H_2O \tag{7}$$

$$O_2^{\cdot-} + HOCl \rightarrow \cdot\mathbf{OH} + Cl^- + {}^1\mathbf{O_2} \tag{8}$$

$$O_2^{\cdot-} + \cdot NO \rightarrow {}^-\mathbf{OONO} \tag{9}$$

All of these secondary ROMs damage cells and eventually kill them. Although the formation of these secondary ROMs has been established *in vitro* (in a test tube), it does not follow that they are also formed in living phagocytes. For decades scientists have argued for and against the formation of these secondary ROMs. I have already discussed some of these controversies (Chapter 2). We now know that all of these ROMs are produced by phagocytes, and all contribute to the killing of the foreign intruder. For a short summary see Eberhardt (2000).

The question arises: if the phagocytes produce such a variety of reactive ROMs, how do they survive? They contain high levels of antioxidants (SOD, CAT, GSH-Px, vitamins C and E and taurine), but despite of it, they eventually die. They make the ultimate sacrifice. The accumulation of dead tissue, dead neutrophils and dead macrophages is known as pus The high levels of the enzyme myeloperoxidase (MPO) gives pus its green color. The fate of activated phagocytes should give us food for thought. These simple organisms die, despite their high levels of antioxidants. We should, therefore, not be surprised to find that in large scale epidemiological studies on humans, antioxidant supplementation has shown less than dramatic effects.

Genetic abnormalities

In biology the study of abnormalities help us understand the normal. A genetic defect in phagocytes is **chronic granulo-matous disease** (CGD). In this condition phagocytes cannot manufacture $O_2 \cdot^-$. This may be due to a defective NADPH oxidase or maybe the enzyme cannot be activated. The patients suffer frequent and protracted infections and usually die in infancy.

Another abnormality is the **myeloperoxidase (MPO) deficiency**. Patients with this condition are usually healthy and can lead a normal life. MPO is involved in the formation of HOCl (reaction 2). Reaction (2) is followed by reactions (7) and (8). Any damage to the enzyme involved in the formation of primary ROMs is of course more damaging than a defective enzyme involved in secondary ROM formation. Without NADPH oxidase we do not produce any ROMs, but without or a damaged MPO we still produce ROMs via reactions (4)-(6) and (9). *Nature* never puts all her eggs in one basket. Whenever one pathway is blocked, we always have a backup system.

Tissue damage by phagocytes

In phagocytosis the ROMs are not completely contained in the phagosome, but some leak out into the extracellular space, causing damage to the surrounding healthy tissue. This is especially true in the case of **failed phagocytosis**. When the phagocyte attacks an inorganic material, which is indigestible, such as an asbestos fiber, the ROMs are continuously produced and released into the extracellular space. Asbestos fibers may be considered to act like an implanted radiation source, producing ROMs and leading to scar tissue formation (**asbestosis**). The damage to external tissue is mostly localized in bacterial infection, but is spread over a larger domain in viral infection.

Since phagocytes are the most prolific producers of ROMs, they initiate many pathological processes, such as **cancer, neurodegenerative disorders and septic shock**. I shall discuss these processes in Chapter 5.

86

Chapter 4

ANTIOXIDANTS

Introduction

If all of these ROMs are so bad for us, why do we still exist? During the course of evolution our bodies have developed defenses to remove these toxic byproducts of our oxygen metabolism. The antioxidant defenses come in two varieties: 1. Enzymes and some peptides and other small molecules produced by our own cells. These are superoxide dismutase (SOD), catalase (CAT), glutathione peroxidase (GSH-Px, melatonin, uric acid, carnosine, glutathione, metal-complexing proteins, etc. and 2. Compounds taken up in our diet, like flavonoids and the vitamins A, C and E.

Antioxidants are essential for the survival of all aerobic organisms. Anaerobic bacteria are killed by oxygen. The enzymatic defenses are only effective over a limited range of oxygen concentrations. Normal air consists of 21% oxygen. Exposure of humans to pure (100 %) oxygen for only a short time causes many deleterious effects and exposure over prolonged periods leads to alveolar damage. This effect of increased oxygen concentration varies considerably depending on the organism, age, physiological condition and diet (initial conditions).

The term antioxidant has become a new catch word in the popular literature. It usually refers to certain vitamins in our diet (vitamins A, C, and E). In the scientific realm however the term 'antioxidant' encompasses a much greater variety of compounds. The most important antioxidants are produced by our own metabolism. This is not surprising, considering that they are the products of millions of years of evolution.

Adaptation to changing conditions is an evolutionary process, taking millions of years. During evolution our bodies have developed antioxidant defenses (SOD, CAT, GSH-Px, uric acid, melatonin, peptides, carnosine) against the reactive oxygen metabolites. However, these defenses are not always sufficient, since we in our fast changing world, are exposed to increased "oxidative stress" from many sources. I shall discuss in detail the concept of "oxidative stress" in Chapter 5. We can increase to some extent our defenses by taking antioxidants in our diet (vitamins A, C, and E, and selenium). Vitamins are compounds, which our bodies cannot synthesize, but which are essential for many biological functions. However it is not clear whether these vitamins offer any additional protection against ROMs, since we already have plenty of antioxidants available. As I shall point out later in this chapter the vitamins C and E can under certain conditions act as prooxidants.

The hydroxyl radical is highly reactive towards most molecules. Therefore most molecules can be considered scavengers of hydroxyl radicals especially electron-rich aromatics (such as the bases of DNA). If we want to protect a biomolecule B from hydroxyl radical attack, we have to add a radical scavenger S, which can compete with the biomolecule:

$$\cdot OH \; + \; B \; \rightarrow \; B\cdot$$

$$\cdot OH \; + \; S \; \rightarrow \; S\cdot$$

where B can be an **enzyme, lipid or DNA**, and S can be any number of molecules, such as ascorbic acid, α-tocopherol, uric acid, flavonoids, carnosine, melatonin and others. The scavenger can compete only at high concentrations, which is physiologically impossible (theoretically it should be possible to scavenge all $\cdot$OH radicals by high concentrations of sugar or alcohol!). In addition the hydroxyl radical is formed in close proximity to the target (site-specific attack on DNA and proteins). In other words we cannot protect our cells and tissues against hydroxyl radicals. However we can prevent hydroxyl radical formation by eliminating the precursors of hydroxyl namely the hydrogen peroxide and metal ions (Fe^{2+}, Cu^+). This was indeed the strategy employed by evolution. The hydrogen peroxide is eliminated by **catalase and glutathione peroxidase** and the metal ions can be sequestered by metal-complexing proteins, like **ferritin, transferrin, ceruloplasmin:**

$$2\,O_2{}^{\cdot-} \xrightarrow[\text{SOD}]{2\,H^+} H_2O_2 + O_2$$

$$2\,H_2O_2 \xrightarrow[\text{GSH-Px}]{\text{CAT}} H_2O + O_2$$

$$Fe^{2+}(Cu^+) + H_2O_2 \rightarrow Fe^{3+}(Cu^{2+}) + OH^- + \cdot OH$$

ferritin $\downarrow$

$$Fe^{3+}\text{-ferritin complex} \xrightarrow[\substack{\cdot NO \\ \text{polyphenols}}]{O_2{}^{\cdot-}} Fe^{2+}\text{-release}$$

However the superoxide radical anion and nitric oxide as well as some other compounds can cause the release of free iron ions from these iron storage proteins. The superoxide radical anion provides in this way the catalyst for the formation of the more reactive secondary metabolites.

The study of antioxidant defenses presents an excellent example of the **pros and cons** of interfering with a complex dynamic system. Oxygen is necessary for survival, but too much oxygen is toxic. A deleterious effect of too little oxygen was also observed. Fibroblasts in tissue culture intermittently exposed to an atmosphere of nitrogen (anaerobiosis) underwent neoplastic transformation. A certain amount of ascorbic acid is necessary to avoid scorbut and as a cofactor for prolyl and lysyl hydroxylases. These enzymes are important in the biosynthesis of collagen. Elevated doses may be beneficial by scavenging oxygen radicals. Ascorbic acid in moderate amounts (250 mg/day) protected DNA in human sperm cells from oxidation (formation of 8-OHdG decreased considerably). Ascorbic acid also inhibits carcinogenic **N-nitrosamine** formation in the stomach, but high doses can lead to many damaging effects, especially in presence of iron and copper ions. These ions can be released from their storage proteins by trauma or by some ROMs ($\cdot NO$, $O_2 \cdot^-$).

For our bodies to synthesize these antioxidant enzymes we need certain compounds. The SODs come in several varieties, one contains **Cu (II)** and **Zn (II)**, while others contain **Mn (III)** or **Fe (III)** at the catalytic center. The enzyme catalase contains **Cu (II)** and glutathione peroxidase contains **selenium (Se)**. These ions have to be supplied by our diet. In a large population study in China where the levels of selenium in the soil are extremely low, a higher incidence of different types of cancer was reported. A certain amount of Cu (II) and Fe (III) are essential for the synthesis of SOD, CAT and GSH-Px, but excess may cause many deleterious effects via formation of the damaging hydroxyl radical.

The SOD paradox

> *... straight lines evidently belong to geometry,*
> *not to nature and life.*
>
> *Hermann Hesse in the Glass Bead Game*

It has been clearly established that $O_2{}^{\cdot-}$ is formed in all aerobic cells and all aerobic cells contain **superoxide dismutase (SOD)** as a defense against superoxide toxicity (Fridovich, 1995). The function of SODs was first illuminated by McCord and Fridovich (1968, 1969). The SODs come in two varieties: 1. SODs containing Cu (II) and Zn (II), and 2. SODs containing Mn(III) or Fe(III) at the catalytic center. SODs catalyze the following reaction:

$$2\,O_2{}^{\cdot-} \;\xrightarrow[\text{SOD}]{2\,H^+}\; H_2O_2 \;+\; O_2$$

In biology the study of the abnormal helps us to understand the normal. E. coli mutants, unable to produce the MnSOD (SodA) or the FeSOD (SodB) enzymes have shown a hyper-sensitivity towards oxygen. These E. coli mutants show enhanced oxygen-dependent mutagenesis.

From these results we may be tempted to conclude: the more SOD the better! However the study of **SOD-rich** *E. coli* bacteria led to a paradoxical increase in $O_2{}^{\cdot-}$ toxicity. Since SOD converts $O_2{}^{\cdot-}$ to H_2O_2 it was concluded that the increase in H_2O_2 (the precursor of the highly destructive $\cdot OH$) accounts for the increased sensitivity of SOD-rich bacteria. However, other investigators came up with a different explanation. Superoxide induces some enzymes of *E. coli*. Overproduction of SOD lowers the steady state concentration of $O_2{}^{\cdot-}$ and thus suppresses the induction of these enzymes (glutathione reductase GSH-red.). These enzymes are involved in the reduction of glutathione disulfide to glutathione:

$$\text{GSSG} \quad \xrightarrow{\text{Glu-red.}} \quad 2\,\text{GSH}$$

SOD lowers the concentration of $O_2\cdot^-$ and consequently decreases GSH, one of the important **antioxidant defenses**.

Whenever we have increased production of $O_2\cdot^-$ our cells respond with increased antioxidant production. I shall come back to this defense mechanism later in Chapter 5 (on exercise). One important damaging reaction of superoxide radical anion is lipid peroxidation. This reaction has to compete with the removal of $O_2\cdot^-$ by SOD:

$$\text{Lipid peroxidation} \xleftarrow{\text{LH}} O_2\cdot^- \xrightarrow{\text{SOD}} H_2O_2 + O_2$$

Work carried out by McCord and coworkers (1994) on lipid peroxidation in **ischemia/reperfusion injury** (Chapter 5) in rabbit heart has provided us with a beautiful example of the dual character of $O_2\cdot^-$.

Lipid peroxidation proceeds as follows:

$$\text{L-H} + \cdot\text{X} \rightarrow \text{L}\cdot + \text{XH} \qquad \text{initiation} \qquad (1)$$

where $X = O_2\cdot^-, \cdot OH, {}^-OONO, \cdot NO_2$

$$\text{L}\cdot + O_2 \rightarrow \text{L-OO}\cdot \qquad \text{propagation} \quad (2)$$

$$\text{L-OO}\cdot + \text{L-H} \rightarrow \text{LOOH} + \text{L}\cdot \qquad \text{propagation} \quad (3)$$

$$\text{L-OO}\cdot + \text{L-OO}\cdot \rightarrow \text{LOOOOL} \qquad \text{termination} \quad (4)$$

$$LOO\cdot + O_2{}^{\cdot-} + H^+ \rightarrow LOOH + O_2 \quad \text{termination} \quad (5)$$

$O_2{}^{\cdot-}$ can act as an **electron acceptor** or an **electron donor.** In lipid peroxidation it can act as a chain initiator (reaction 1), but also as a chain terminator (reaction 5). Since SOD removes $O_2{}^{\cdot-}$ it not only lowers the rate of initiation, but also lowers the rate of termination. In other words SOD can be both beneficial and detrimental, too much of a good thing can be bad. This dual behaviour of SOD has become known as the **SOD paradox.** There are several neurodegenerative diseases (amyotrophic lateral sclerosis, Alzheimer's, Down's Syndrome and Parkinson's disease) in which an imbalance of SOD is involved.

Normally the antioxidant defenses are sufficient to keep us from serious harm. We have a balance between the prooxidant and antioxidant forces. Whenever this balance is disturbed, as is the case in many pathological conditions, we have a situation called **"oxidative stress"**.

The important aspect is the proper prooxidant/antioxidant balance. Whenever this balance is disturbed medical science tries to fix the problem by adding another chemical (drug) to an already complex mix. The results can be catastrophic (remember thalidomide). All prescription drugs contain a litany of possible side effects, which may or may not occur. We are dealing with a complex dynamic system, where the initial conditions are different for each individual. This is why medicine is an art and not yet a science!

Small molecules produced via our own metabolism

The most important antioxidant of small molecular weight produced by our own metabolism is **uric acid**. The reaction of a number of reactive oxygen species with uric acid was found by Ames et al.(1981) to produce allantoin, as the major product:

$$uric\ acid \xrightarrow[\cdot OH,\ ^1O_2]{^-OONO} allantoin$$

It was, therefore, suggested by these authors that uric acid is an *in vivo* antioxidant and the determination of allantoin may serve as a marker for oxidative stress *in vivo*. Support for this idea comes from patients with rheumatoid arthritis. In these patients the level of allantoin was higher and the level of uric acid was lower compared to normal subjects.

Uric acid does not react with ·NO and therefore does not inhibit the many important physiological functions of ·NO. Uric acid only scavenges the highly damaging nitric oxide deactivation product, the peroxynitrite. Uric acid is therefore an evolutionary success story: it does not affect the physiological functions of ·NO, but only protects cells from its toxic effects. All the enzymatic antioxidants (SOD, CAT, GSH-Px) are catalysts, they are not consumed contrary to the small molecular weight antioxidants (such as uric acid, melatonin, the vitamins A, C and E), which protect by **self-sacrifice.** In all of these scavenger reactions the uric acid is consumed. However there is another way by which uric acid can act as an antioxidant, without being oxidized itself. This process is complexation of metal ions, such as Fe^{2+} or Fe^{3+}. Uric acid can inhibit the iron-catalyzed lipid peroxidation by complexation of the iron ions. This complexing ability of uric acid is most likely responsible for the inhibition of ascorbate autoxidation in human serum.

Uric acid is indeed the most important antioxidant in human plasma. The urate level is considerably higher (~300 μM) than the ascorbate level (~50 μM). The increase in urate levels occured during the course of 60 million years of evolution and coincided with a large increase in life-span and brain size. At the same time an enormous decrease in the age-specific cancer rate has occured in humans compared to short-lived mammals.

The lower life expectancy of rats has been suggested to be due to the higher metabolic rate of rats (higher oxygen consumption) and consequently formation of higher concentrations of reactive oxygen metabolites. Rats excreted much more 8-OHdG (a measure for *in vivo* oxidative damage) in the urine than humans. However, as suggested by Ames et al (1981) we may also surmise that the longer lifespan is due to better antioxidant defenses. The uric acid concentration in humans is considerably higher (5 mg/100 ml of blood) than in rats (0.5 mg/100 ml of blood). The longer life expectancy of humans is most likely due to a combination of these factors (less oxidative stress and better antioxidant defenses).

If we exercise we use more oxygen and produce more reactive oxygen metabolites, as shown by increased lipid peroxidation. This effect can be determined *in vivo* by measuring hydrocarbons in the exhaled breath (Riely et al. 1974, Muller and Sies, 1984). At the same time urate levels in the blood increase possibly as a physiological response to this increased oxidative stress. Exercise also increases the amount of some small peptides, such as carnosine, homocarnosine and anserine in the muscles of rowers and sprinters. These defensive responses are unlikely to be perfect, and we may therefore ask: Is exercise good or bad for you? This question is difficult to answer, since we are dealing with a complex dynamic system where the initial conditions are different for each individual. Evolution has replaced ascorbic acid with uric acid, which is a more effective antioxidant with fewer of the damaging side effects of Vitamin C, which we have already discussed. It took evolution millions of years to adopt the best strategy against oxidative stress. In light of this, the attempt to improve on this strategy by recommending high doses of Vitamin C appears futile at best.

Since uric acid is an antioxidant, we may assume, that the higher the level of uric acid in our plasma, the better. However remember the words of Paracelsus: alles is Gift, es kommt nur auf die Dosis an. Too much uric acid leads to the development of **gout.** On the other hand, large scale epidemiological studies have shown, that people with gout never contract **multiple sclerosis.** It is clearly a question of **balance.**

Another interesting aspect of uric acid is the observation that a high protein diet (which leads to high uric acid levels) lowers the life expectancy in rodents and mammals. Uric acid acts like Dr. Jekyll and Mr. Hyde. At low levels uric acid is beneficial, but at higher concentrations, it leads to gout, gall and kidney stones, and lowers the life expectancy.

Another important small molecular weight compound, produced by our own metabolism is N-acetyl-5-methoxy-tryptamine or better known as **melatonin.** Melatonin is a hormone, which for many years was thought to be produced exclusively by the pineal gland. However it has been shown that melatonin is produced in vertebrates in a number of cells and organs. It has the following structure:

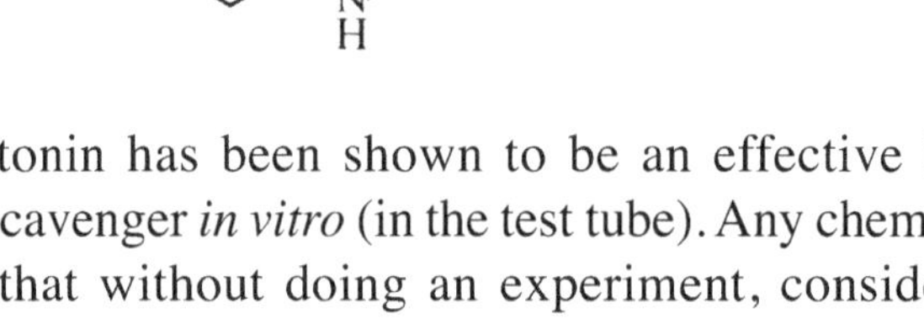

Melatonin has been shown to be an effective hydroxyl radical scavenger *in vitro* (in the test tube). Any chemist would predict that without doing an experiment, considering the electron-rich aromatic ring and the highly electrophilic hydroxyl radical. However, if melatonin indeed acts as an antioxidant *in vivo* (in a living organism), it must be able to penetrate cells and subcellular compartments where the ROMs are generated. Since melatonin is highly lipophilic, it meets this requirement admirably. Melatonin, contrary to Vitamins C and E does not act as a prooxidant (does not produce radicals).

Melatonin does not scavenge the primary metabolite superoxide radical anion ($O_2 \cdot^-$), but scavenges the secondary metabolites singlet oxygen ($^1O_2$) and the hydroxyl radical ($\cdot OH$). Melatonin in scavenging HOCl also protects CAT from inactivation by HOCl. Each of the ROMs (**HOCl, $^1O_2$, $\cdot OH$**) has several possibilities to react: they can either damage important biomolecules (DNA, proteins, lipids, CAT) or they can be scavenged by the antioxidant melatonin. We always have competition between several reactions and therefore the antioxidant can never offer complete protection.

Protective effects of melatonin on oxidative damage *in vivo* has been demonstrated in several animal models. Chemical-induced injury to lung and liver of rats was significantly reduced by co-treatment with melatonin. Safrole, an extract of sassafras oil, is widely used as a cancer initiator. Rats injected with safrole showed extensive damage to their liver DNA. Co-treatment with melatonin significantly reduced the DNA damage.

In addition melatonin acts as **biological clock**. It regulates the day-night rhythm. Blood levels of melatonin are low during the day and high at night (Reiter and Robinson 1995). At high nocturnal levels of melatonin safrole induced less DNA damage than at low melatonin levels during the day. This observation may have nothing to do with melatonin. At night we have a lower metabolic rate (lower O_2 consumption) and consequently a lower rate of ROM formation and ROM-induced damage.

Melatonin has been hailed in the Press and in some best-selling books as a new wonder drug (Reiter and Robinson 1995). Melatonin is recommended as a sleeping pill, for prevention of cancer, heart disease, Alzheimer's disease, AIDS, depression and old age. Since it is promoted as a dietary supplement, it is a non-prescription drug sold in health food stores. Melatonin has never been tested for safety in humans. Most scientists agree that the 3 mg tablets available in health food stores is too high a dose to restore a normal sleeping pattern. Some researchers report that a dose of 0.3 mg completely restored sleep to normal, and both 0.1 mg and 3.0 mg. was less effective. More is less, and less is more! The optimum dose to restore a normal sleeping pattern may be quite different from the dose needed to prevent cancer, heart disease, Alzheimer's disease and aging.

For the synthesis of melatonin our bodies need **tryptophan** and **calcium (Ca²⁺)**. Supplementation by these nutrients has indeed shown an increased production of melatonin (Reiter and Robinson 1995). Grandma's prescription for a good nights sleep of a warm glass of milk (contains tryptophan and Ca²⁺) may therefore have a scientific basis. The biosynthesis of melatonin can be inhibited by many common compounds, among them **aspirin, caffeine and Ca²⁺- channel blockers** (heart medications) (Reiter and Robinson 1995). Patients who take a daily dose of aspirin to prevent coronary artery disease, should, therefore, not take it at night, but in the morning.

Dietary antioxidants

Ascorbic acid (AH_2) or Vitamin C is a very important water-soluble radical scavenger or antioxidant. Ascorbic acid has the following structure:

$$AH_2 \qquad\qquad A$$

It is evident from the structure that ascorbic acid has two easily abstractable H-atoms (the two OH groups on the five-membered ring).The most important property of ascorbic acid is its ability to be oxidized to dehydroascorbic acid (A). This transformation represents a two-electron oxidation, which proceeds in two 1-electron steps:

$$AH_2 \;\rightarrow\; AH\cdot \;\rightarrow\; A$$

The transformation of ascorbic acid to dehydroascorbic acid of course destroys its effectiveness as a vitamin and antioxidant. The mechanism of ascorbic acid oxidation by oxygen was therefore studied extensively. Metal ions can facilitate the transfer of an electron from ascorbate to oxygen. Therefore, orange juice should not be stored in a metal container.

In addition to the oxidation of ascorbic acid by oxygen, superoxide and perhydroxyl radicals, ascorbic acid is oxidized by hydroxyl radicals. The rate constant for this reaction is faster by several orders of magnitude than the aforementioned oxidations. The reaction is diffusion controlled ($k=1.2 \times 10^{10}$ M^{-1}s^{-1}), i.e. the ·OH reacts at every collision:

$$AH_2 \;+\; \cdot OH \;\rightarrow\; AH\cdot \;+\; H_2O$$

$$AH^- + \cdot OH \rightarrow A^{\cdot -} + H_2O$$

Some animals are able to synthesize ascorbic acid, but humans have lost this ability during evolution. The antioxidant preserved through millions of years of evolution is the uric acid, which has replaced ascorbic acid as the most important antioxidant in human plasma. It would be foolhardy to assume that evolution has chosen the wrong path. As I have already pointed out, uric acid is an evolutionary success story. It reacts with the damaging metabolite of NO (the ^-OONO), but does not react with NO and thus does not interfere with the physiological function of NO.

Vitamin C was plentiful in the food supply during human evolution. Therefore, it was easier for man to obtain Vitamin C through the food, rather than use energy and nutrients for its synthesis. We find the same situation with other vitamins, such as Vitamin A. It cannot be synthesized by the cat. Since Vitamin A is important for night vision, cats have to obtain Vitamin A through their prey, which have accumulated it in their livers. Since cats just like humans no longer live in a natural habitat, the foods have to be supplemented with these vitamins. The question is: how much? Remember Paracelsus: Everything is poison, it just depends on the dose.

The pros and cons of ascorbic acid

There is no doubt that ascorbic acid can act as an anti-oxidant. It reacts with all the ROMs. However ascorbic acid is also a **reducing agent** (i. e. it donates an electron) and therefore reacts with electron-accepting metal ions, such as Fe^{3+} or Cu^{2+}, written in simplified form:

$$Fe^{3+} + AH^- \rightarrow Fe^{2+} + AH\cdot$$

$$Cu^{2+} + AH^- \rightarrow Cu^+ + AH\cdot$$

Since the reduced metal ions are the essential catalysts for the formation of the highly destructive hydroxyl radical (Chapter 2) ascorbic acid acts as a **prooxidant.** via the above reactions. Based on this prooxidant activity of Vitamin C, B. Halliwell (1994) published an interesting article with the provocative title:"Vitamin C, the key to health or a slow-acting carcinogen"?

This reducing ability of ascorbate can not only be deleterious, but can also have a beneficial effect. It prevents the formation of N-nitrosamines in the stomach. We have the following competing reactions:

$$R_2\text{-NH} \quad + \text{HONO} \quad \rightarrow \quad R_2\text{-N-NO}$$

$$AH_2 + \text{HONO} \quad \rightarrow \quad AH\cdot + H_2O + \cdot NO$$

Ascorbic acid competes with secondary amines for nitrous acid (HONO), and thus prevents formation of the highly **carcinogenic N-nitrosamines.** The reducing ability of ascorbic acid may also have beneficial physiological effects. Ascorbic acid can regulate the blood supply in the brain. The brain requires a precise means of regulating oxygen delivery to the neurons. Too much or too little oxygen can cause serious damage. Nitric oxide is a vasodilator produced by many different cells including neurons. So the blood supply could possibly be regulated via nitric oxide. However, in hypoxia the synthesis of $\cdot$NO from L-arginine is unlikely, since this synthesis requires oxygen. It has been suggested that ascorbate release from neurons generates $\cdot$NO via reduction of nitrite (NO_2^-) in the extracellular space. This way neurons may regulate their own oxygen supply. The consequences of low ascorbate in the brain would be progressive damage from inaccurate oxygen delivery.

All these prooxidant effects should not detract our attention to the importance of ascorbate as a cofactor for some enzymes and as an antioxidant. Ascorbic acid is a cofactor for some hydroxylase enzymes, the lysyl and prolyl hydroxylases. These hydroxylations play a role in the biosynthesis of collagen. Ascorbic acid accelerates hydroxylation, by donating an electron to the metal-containing enzyme for which the metal ion is essential for optimal activity.

So the biochemically important role of ascorbate is to act as an **electron donor.** The property which makes ascorbic acid useful in biochemical synthesis also makes it a possibly dangerous drug via formation of $O_2 \cdot^-$, H_2O_2 and $\cdot OH$.

I have already mentioned the electron donating reaction of lipid hydroperoxides with Vitamin C. This reaction gives alkoxyl radicals and damages DNA (p. 75).

As with other antioxidant defenses, such as SOD, CAT, GSH-Px we find higher concentrations of ascorbate in organs which are more exposed to oxidative stress due to higher metabolic rate and oxygen consumption. As I have pointed out above an adequate oxygen supply is very important for the brain. The human brain makes up only about 2% of the total body weight, but consumes about 18% of the body's total oxygen consumption. The pros and cons of Vitamin C supplementation have been excellently reviewed by Halliwell (1994), who states, and I quote: 'Hence there is no clear evidence for any great benefit to be obtained by mega-doses of Vitamin C, and we cannot yet prove that it is not harmful over a lifetime'.

As I mentioned before antioxidant vitamins protect through **self-sacrifice**, resulting in a low level of these antioxidants in the plasma of patients under oxidative stress. Therefore, the FDA has recently increased the RDA (recommended daily allowance) of Vitamin C for smokers (see Drug Information 2001 of the American Society of Health-System Pharmacists) from 100 mg/day to 125 mg./day. The RDA for healthy males is 90 mg/day, and for females 75 mg/day. The increased RDA for smokers may give the mistaken and dangerous impression, that Vitamin C protects you from the damaging effects of smoking. Antioxidant vitamins do not protect you from lung cancer and β-carotene may even be harmful (Chapter 6).

Next to ascorbic acid, **Vitamin E** represents the most important antioxidant. Vitamin C and E complement each other. Ascorbic acid is water soluble and Vitamin E is lipid soluble and is incorporated into the lipid bilayer, where it is needed most. Vitamin E deficiency in old chicken feed (high in polyunsaturated fats) resulted in the destruction of the cerebellum of the chicken. This was called 'crazy chick' disease. From this observation we may conclude that without Vitamin E complex brains could not have developed. In other words: evolution was pushed in the direction of more complex brains, because of the available nutrient, Vitamin E and its chemistry. Without the availability of Vitamin E in our food supply, we would not have evolved to the level of *homo sapiens*. A detailed account of food as an evolutionary instrument is given by Crawford and Marsh in 'Nutrition and Evolution' (1995). The presence or absence of chemicals, such as oxygen, or nutrients like Vitamin A or E make certain evolutionary directions possible or impossible.

The important balance between Vitamin E and oxygen metabolism is best illustrated by the tragic example of premature babies. These babies were placed in an oxygen tent and became blind, because the mothers did not transfer enough Vitamin E to their fetuses. From tragic events like these we sometimes learn something, so in the end some good will result. Remember: the spots in the Yin-Yang symbol (even the bad has some good in it). The Puerto Ricans have a saying: *no hay mal, que por bien no venga.*

Vitamin E is a mixture of tocopherols of which the **α-tocopherol** (α- TOH) is the most powerful natural inhibitor of lipid peroxidation by terminating the chain oxidation:

$$LOO\cdot \ + \ \alpha\text{-TOH} \ \xrightarrow{k_t} \ LOOH \ + \ \alpha\text{-TO}\cdot$$

The α-tocopheroxyl radical (α-TO·) is relatively stable. We have again an example in which a highly reactive radical is transformed to a less reactive one.

The α-TO· may under certain conditions (low density lipoprotein or LDL peroxidation) continue the chain oxidation:

$$\alpha\text{-TO}\cdot \ + \ LH \ \xrightarrow{k_{init}} \ \alpha\text{-TOH} \ + \ L\cdot \qquad (1)$$

$$L\cdot \ + \ O_2 \ \rightarrow \ LOO\cdot \qquad (2)$$

$$LOO\cdot \ + \ \alpha\text{-TOH} \ \rightarrow \ LOOH \ + \ \alpha\text{-TO}\cdot \qquad (3)$$

This sequence of reactions, the tocopherol-mediated peroxidation (TMP) endows tocopherol with a **prooxidant activity**. The complex kinetics involved in the α-TOH-mediated peroxidation (TMP) of low-density lipoprotein (LDL) particles have been investigated , but are beyond the scope of the present booklet (Bowry and Ingold, 1999). These reactions are involved in the initiation of **atherosclerosis**.

Another important reaction of the α-TO· is the regeneration of α-TOH by its reaction with co-antioxidants, such as ascorbate:

$$\alpha\text{-TO}\cdot \ + \ AH^- \ \xrightarrow{k} \ \alpha\text{-TOH} \ + A\cdot^-$$

Both Vitamin C and Vitamin E are chain-breaking antioxidants, and there is extensive and conclusive evidence, which indicates that there is a synergistic antioxidant interaction between these vitamins in a variety of *in vitro* systems. In other words we have cooperation between these two antioxidant vitamins.

Another way by which α-TOH may act as **a prooxidant** is by reduction of **transition metal ions** (Proudfoot et al. 1997):

$$\alpha\text{-TOH} \; + \; Cu^{2+} \; \rightarrow \; \alpha\text{-TO·} \; + H^+ \; + \; Cu^+$$

followed by :

$$LOOH \; + \; Cu^+ \; \rightarrow \; LO· \; + \; OH^- \; + \; Cu^{2+}$$

$$LO· \; + \; LH \; \rightarrow \; LOH \; + \; L·$$

These reactions are analogous to the ascorbic acid case. We have again a pro and con situation. The prooxidant/antioxidant activity of α-TOH and the role of Cu^{2+} in the oxidative modification of **low density lipoprotein (LDL)** is involved in the first step of **atherogenesis,**

The **carotenoids** comprise a group of natural products present in many fruits and vegetables (such as carrots). Their common chemical characteristic is a long hydrocarbon chain with conjugated double bonds:

β-carotene

These conjugated chains give the carotenoids their characteristic color (yellow-red).

β-carotene is converted via enzymatic reactions to two molecules of **Vitamin A** (retinol). Vitamin A is important for vision, especially night vision. Vitamin A deficiency causes blindness. The ability to form Vitamin A from carotene is lost in cats, which need it for night vision. Cats used to obtain plenty of Vitamin A from the livers of their prey. However since cats, like humans, no longer live in a natural environment (cats get most of their food from their owners), their food has to be fortified with Vitamin A. Cases of Vitamin A deficiency are rare in the USA , but common in developing nations. A recent report by a panel of the Institute of Medicine (Robert Russell) was more concerned with the maximum daily dose, which they set at 3000 micrograms/day. Many multivitamins available in Health Food stores contain more than this upper limit (up to 8000 micrograms). Excess Vitamin A is toxic, leading to liver damage, birth defects and easily fractured bones (**pro-con**). Since excess Vitamin A is stored in the liver, it does not have to be taken on a daily basis in order to avoid Vitamin A deficiency. This is contrary to Vitamin C, which is water soluble and any excess is excreted in the urine. Vitamin C must therefore be replenished on a daily basis.

The carotenoids are best known for their ability to quench **singlet oxygen:**

$$\text{damage to biomolecules} \quad \leftarrow \quad {}^{1}O_2 \quad \xrightarrow{\text{carotenes}} \quad {}^{3}O_2$$

One example of this quenching effect is the treatment of **erythropoietic porphyria**, a human photosensitivity disease. The skin of these patients accumulates high concentrations of porphyrins, which react with oxygen to give singlet oxygen via photosensitization (Chapter 2). The symptoms of these patients can be relieved by β-carotene (Mathews-Roth et al. 1970). This result indicates that singlet oxygen is involved in causing the symptoms of this disease, but not causing the underlying pathology (**cause and effect**).

The most effective singlet oxygen quencher among the carotenoids is **lycopene**. The β-carotene is only half as effective. Other antioxidants, like ascorbic acid and the tocopherols also quench singlet oxygen, but at rates many orders of magnitude smaller than the carotenoids. Lycopene has recently received a lot of attention, because studies have shown a relationship between diets rich in lycopene and a reduced risk of prostate cancer. Tomatoes are the most important source of lycopene. Other sources are pink grapefruit and watermelon.

In addition to quenching singlet oxygen, carotenes act as radical scavengers:

$$\text{damage to biomolecules} \quad \leftarrow \quad \begin{bmatrix} ROO \cdot \\ O_2 \cdot^- \\ HO_2 \cdot \\ HO \cdot \end{bmatrix} \quad \rightarrow \quad \text{scavenged by carotene}$$

Another interesting effect was ascribed to the scavenging of singlet oxygen. Human **polymorphonuclear leukocytes** (PMNs) killed a colorless mutant of *Sarcina lutea* much more readily than the carotenoid-containing strain (Krinsky, 1974). Since carotenoids are singlet oxygen quenchers, the protective effect was considered as evidence of singlet oxygen formation in the bactericidal action of leukocytes. PMNs however also produce radicals ($\cdot$OH), so the observed effect could be due to scavenging of radicals.

Flavonoids

a glass of red wine a day keeps the doctor away

Our daily diet not only contains vitamins, but also other natural products, which act as antioxidants. Some of these products are the **flavonoids,** which are abundant in fruits, vegetables, tea and red wine. Chemically, the flavonoids contain phenolic or polyphenolic groups (easily abstractable H-atoms). Flavonoids are, therefore, efficient scavengers of reactive oxygen metabolites, and thus protect our membranes from lipid peroxidation, by the same mechanism as Vitamin E. However like Vitamin E flavonoids can also act under certain conditions as **prooxidants** (Laughton et al., 1989, Rahman et al., 1989). In these cases flavonoids act as reducing agents for Fe^{3+} or Cu^{2+} to give Fe^{2+} or Cu^+, which are the essential catalysts for the formation of the hydroxyl radical (Chapter 2). Therefore *in vivo* the levels of these free metal ions must be carefully controlled. The metals are complexed to a number of high molecular weight proteins (ferritin, ceruloplasmin, albumin).

Atherosclerosis has long been associated with a high fat diet. French people eat a lot of high fat food (cheeses), but have low mortality rates from coronary artery disease. This observation has become known as the **French paradox**. How can we explain this paradox? Possibly because they drink lots of red wine, which is rich in antioxidant flavonoids. On the other hand the French have a higher incidence of liver cirrhosis. There is a point where the damaging effects of alcohol outweigh the beneficial effects of flavonoids. It is a question of **balance**.

In addition to flavonoids, our foods contain many other phenolic and polyphenolic compounds, which are added to prevent spoilage (lipid peroxidation). One such compound is tannic acid. These compounds may act as anticarcinogens, but also as mutagens and procarcinogens (Ames, 1983, Stich, 1991). The combination tannic acid-oxygen-Cu^{2+} has been shown to cause DNA strand breaks *in vitro*. It is again a question of **balance.**

110

Chapter 5.

OXIDATIVE STRESS AND PATHOLOGICAL CONSEQUENCES

" For like every great idea it has no real beginning; rather, it has always been there, at least the idea of it."
Hermann Hesse in "The Glass Bead Game"*

Introduction

Our metabolism produces a constant flux of reactive oxygen metabolites, which attack membranes (lipids), enzymes (proteins) and DNA. I have already discussed how these ROMs react with these biomolecules. There is nothing we can do to prevent their formation, since they are the by-products of our own power plant (the mitochondria) and a direct consequence of life in an oxygen atmosphere. In addition ROMs are formed in microsomes, by endothelial cells, neurons and most of all in very high amounts by phagocytes.

However our cells have evolved **antioxidant defenses** (Chapter 4). Without them we would not have evolved to the level of *homo sapiens*. The ROMs are formed and consumed (by antioxidants) and are therefore present in a steady state concentration (the rate of formation is equal to the rate of consumption). This steady state is usually low enough to cause no major damage.

If the **prooxidant/antioxidant balance** is disturbed, either by a change in the prooxidant forces or a change in the antioxidant forces, we have a situation called **"oxidative stress"** (Sies, 1986). Continued oxidative stress leads to cellular damage and pathological changes. The major factor in pathological changes are the attack on **lipids, enzymes and DNA** The change in the prooxidant/antioxidant balance can be brought about by some abnormality in the metabolism, which can have many causes, like genetic defects (ALS, Down's syndrome, multiple sclerosis) or can be brought about by external factors. Some of these factors can be exposure to viruses, toxic chemicals or improper lifestyle (smoking, alcohol consumption, exercise, diet).

The importance of **balance** in life has been known for centuries by Chinese philosophers. These philosophers emphasized the balance between **Yin and Yang**, the male and female, day and night, spring (planting) and fall (harvesting). The Yin-Yang symbol, as shown on the back cover represents this balance. The black dot in the white part and the white dot in the black part represent the idea that each time one of the two forces reaches its extreme, it contains in itself already the seeds of its opposite. Too much Yang can act as Yin. We encounter the same idea in ROM chemistry. Too little or too much ascorbic acid is bad (it can act as a prooxidant or antioxidant). Too little or too much $O_2 \cdot^-$ or SOD can be bad (**SOD paradox**). I have already mentioned the famous statement by **Paracelsus**: Alles ist Gift, es kommt nur auf die Dosis an. The idea of balance in *Nature* was already recognized by **Aristotle** (the golden mean). It has been said, that there is no human thought, which has not been thought before. However it is up to science to prove it.

We cannot shut off the power plant, but we can lower its power output by lowering our caloric intake. **Caloric restriction** has indeed been shown to increase life expectancy and lower the risk for numerous diseases, such as cancer and heart disease (Tannenbaum, 1942, Masoro et al., 1982, Weindruch et al., 1986, Sohal et al. 1996). It is interesting to point out, that

Tannenbaum made his observations several decades before anybody knew about reactive oxygen metabolites.

Low metabolic rate (low oxygen consumption) also increases life expectancy and decreases the likelihood for pathological processes (Youngman et al., 1992).

Humans have much lower metabolic rates than rodents. At the same time humans have higher levels of antioxidants (such as uric acid). The combination of these two factors are the reason why humans live longer than rats. The high metabolic rate of rats is also responsible why there are so many rodent carcinogens.

Before discussing some specific examples of oxidative stress, we have to ask an important question: are ROMs the cause of the disease or are they the consequence of some underlying pathology?

One of the basic tenets of science is: **every effect has a cause**. Whenever we observe increased formation of ROMs in any pathological process, we have to ask if the increased formation of ROMs is the cause of the pathology or a consequence of it. As I have discussed ROMs react with lipids, destroying the membranes and leading to cell death. We have the two possible sequences of events:

increased formation of ROMs → increased lipid peroxidation → increased tissue damage and cell death

cell death → increased formation of ROMs → increased lipid peroxidation → increased tissue damage and cell death

It is well known that dead cells release metal ions from their storage proteins, and thus accelerate formation of hydroxyl radicals (Chapter 2) and catalyze lipid peroxidation. The observation of reactive oxygen species, either directly or by amelioration of the symptoms by antioxidant therapy, is no proof that the ROMs are causing the pathological abnormality. The cause most likely is some genetic defect such as the over- or under-expression of some genes specifying some enzymes, which change the prooxidant-antioxidant balance.

We read a lot in the Press about all the man-made chemicals in our environment and the danger they pose for our health. We have to realize however, that even in absence of all the chemicals in our external environment would we observe aging, cancer and death. These type of changes (absence of external sources) in our DNA are referred to as **spontaneous mutation** and **spontaneous carcinogenesis** (Smith 1992). The cause of these changes are the ROMs produced by our own metabolism and a direct consequence of living in an oxygen atmosphere.

At present melatonin appears to be the new wonder drug. It has been determined that melatonin levels are much lower in patients with breast cancer, prostate cancer and CAD than in normal patients. The question is: are the low melatonin levels causing the disease or are they the consequence of some pathological process. It is most likely that it is a consequence of increased formation of ROMs, which destroy the antioxidant melatonin.

High concentrations of H_2O_2, $O_2^{\cdot -}$, $\cdot NO$ and ^-OONO are not the causes of amyotrophic lateral sclerosis (ALS), Alzheimer's disease (AD), Parkinson's disease (PD), Down's syndrome (DS) or multiple sclerosis (MS), but arise as a consequence of some metabolic imbalance. The metabolic imbalance may have many causes (genetic defect, hormonal imbalance, loss of homeo-static control, infection, toxic chemicals etc.) and these causes may have other causes all the way down the line to the first cause. Likely candidates for the first cause are radicals which are produced either internally in our bodies (spontaneous mutation, spontaneous carcinogenesis) or can be produced by some toxin (via metabolism to radicals) or some other outside agent like bacteria or viruses. Hydroxyl radicals are formed from the radiation of trace amounts of tritiated water contained in the water of our bodies (Chapter 1). This effect of radiation may have played an important role in our evolution. .

Death of a cell

There are two ways for a cell to die. One way called **necrosis** involves outside agents (toxins), which produce ROMs during their metabolism. ROMs may come from environmental pollution, or may be produced by our own bodies in response to bacteria or viruses or through a number of life style factors, such as diet, smoking or alcohol consumption. ROMs damage the lipids in membranes, thus making the membrane more permeable. These damaged cells permit an increased **influx of Ca^{2+}**, which in turn activate endonucleases. **Endonucleases** are enzymes that break down DNA to individual nucleotides. These nucleotides are then reused for the synthesis of new DNA, and the damaged units (for example 8-OHdG) are excreted in the urine. The determination of **8-OHdG** in the urine is therefore a direct measure of **oxidative stress** in the body.

The other way is known as **apoptosis** or programmed cell death, and is under genetic control (Schulz-Aellen, 1997). The cells do not die due to external factors (toxins), but because death is programmed by the master planner (the DNA). It has often been stated (scientifically incorrect) that non-functional cells commit suicide. The death is preordained by the master planner.

During senescence, there is a reduction in cell proliferative potential and a gradual loss of cells, leading to atrophy and loss of physiological function in most tissues and organs. It is important for an organism to eliminate cells, which do not function properly, just as a company fires an employee, who doesn't do his or her job. This makes good economic sense for an organism, as well as for a superorganism (a company or a nation). An organism can very well tolerate a limited number of non-functional useless cells, but dies if their numbers explode. The same principle applies to an economy with too many welfare recipients. Obviously *Nature* is amoral, guided only by the principle of the survival of the fittest. In this connection it is interesting to note, that economists are increasingly looking to *Nature* as a guide for their models (Pascale et al. 2000).

A decline in the efficiency of the apoptotic program, designed to eliminate damaged cells, would result in an increased number of deleterious, non-functional cells, which might lead to diseases. The manipulation of apoptosis genes could increase life span and decrease the likelihood of diseases of old age (Schulz-Aellen, 1997).

The field of **gerontology** (the science of aging) is a fast evolving field, possibly due to our fast increasing senior generation. Gerontology is studied at numerous levels, at the population level (life span of individuals), at the individual organism level (changes in physiological and biochemical functions), the cellular level and at the subcellular level (changes in molecules, such as lipids, proteins and DNA). The study of gerontolgy, therefore requires the cooperation of many scientist with different backgrounds. Every scientist pushes his or her theory according his or her field of expertise. Theories of aging are therefore a dime a dozen.

In biology the study of the abnormal helps us understand the normal. Genetic diseases that accelerate aging are known as **progerias**. The study of these diseases may help identify the specific gene or genes important for human life span. These hereditary diseases of premature aging have been used as an argument in favor of a genetic basis of aging. Although many of these genetic disorders have shown the same symptoms as in aging, none of the studied diseases has shown all of the aging symptoms. This would indicate that aging is determined by several genes. Most likely aging is a combination of several factors, both **internal and external**. A detailed discussion of different theories of aging is given by Marie-Françoise Schulz-Aellen in *Aging and Human Longevity*. Since the present book is not about aging, but about ROMs, I like to stick to the so-called **"free radical theory of aging"**. The term "free radical" is scientifically incorrect. The adjective "free" is only of historic significance (Chapter 2, p.52). As I have pointed out, in addition to radicals there are many non-radical oxygen metabolites, which cause damage to lipids, proteins and DNA, and thus are involved in aging, and many other diseases. These reactive species include peroxynitrite ($^-$OONO), hydrogen peroxide (H_2O_2), electronically excited states (from dioxetanes) and singlet oxygen ($^1O_2$). The term "free radical theory of aging should be changed to **reactive oxygen metabolite (ROM) theory of aging.**

The theory of aging as originally proposed by Harman in 1981 states, that the accumulated damage to nuclear DNA by radicals is responsible for aging. However, more recent studies have shown, that damage to mitochondrial DNA (the power plant) is the more important factor in aging (Yakes and Van Houten 1997).

Our health and longevity is determined by the following factors: **genes, environment** and **lifestyles.** Since the **prooxidant/antioxidant balance** can be disturbed both by genetic factors, as well as environmental factors (toxins and life styles), it is not surprising that ROMs play an important role in aging, cancer and numerous other diseases. Since we have no control over our genes (at least not yet), our lives can be compared to a card game. We are dealt a certain set of cards and have to play the best game we can. The winner not only depends on the hand dealt (genes), but also on the expertise of the player (lifestyle). Some players play the game of life better than others. I shall discuss the topic *Nature* versus *Nurture* in Chapter 6.

We all recognize aging at the level of individuals, but how do we determine aging at the cellular or molecular level? I have already discussed some molecular markers of ROM-induced damage in Chapter 2. The most commonly marker for aging is **lipofuscin**. Lipofuscin is a complex polymeric mixture, which contains lipids and proteins and exhibits autofluorescence and accumulates in the cytoplasm with age under normal physiological conditions. The fluoroescence makes lipofuscin easy to detect in biological samples. The formation of lipofuscin involves formation of malonaldehyde or other aldehydes formed by lipid peroxidation (Chapter 2). The malonaldehyde interacts with amino groups of proteins to form a Schiff base leading to a complex polymeric material. Since lipofuscin contains protein, it is a good complexing agent for metal ions. Lipofuscin contains **high levels of Cu^{2+} and Fe^{3+}**. Lipofuscin has been characterized as a waste basket for cellular wastes primarily consisting of intracellular membranes.

The ROM theory of aging is supported by studies showing a relationship between oxygen metabolites and lipofuscin accumulation in cultured human glial cells (Thaw et al. 1984). Glial cells are the macrophages of the CNS. These phagocytic cells are the source of copious amounts of reactive oxygen metabolites (Chapter 3). Glial cells were grown in 5%, 10%, 20%, and 40% oxygen. The rate of lipofuscin accumulation increased with increasing oxygen concentration. The presence of the prooxidant combination ascorbic acid - Fe^{3+} increased and the presence of antioxidants such as vitamin E, Se, GSH, DMSO decreased the accumulation of lipofuscin.

Lipofuscin formation is influenced by diet and pathological conditions. The increased formation of lipofuscin in some diseases (Alzheimer's, Parkinson's, amyotrophic lateral sclerosis and multiple sclerosis) will be discussed later in this chapter. Lipofuscin in absence of any pathological condition, is a useful marker for **physiological aging**. Lipofuscin is a marker for aging, but does not cause aging, just as the wrinkles in our face are signs of aging, but not the cause of it.

The mutation-cancer link

Soon after the discovery of X-rays by Roentgen it became evident that repeated X-ray exposure (produces ·OH radicals) correlated with subsequent cancers, such as skin cancers, leukemias and bone cancers.

Herman Muller at Columbia University in 1927 noted that X-irradiation of *Drosophila* fruit flies often resulted in mutant offspring. Timofeéff-Ressovsky, Zimmer and Delbrück in 1935 exposed fruit flies to X-rays and measured the relationship between dose of radiation and the rate of appearance of genetic **mutations**. These observations led to the speculation that the effect of radiation, namely cancer induction and genetic mutation may be directly related. Radiation should, therefore be viewed as a mutagen, capable of damaging the genetic material (DNA) inside cells. Nowadays, we know that radiation of water produces the highly destructive hydroxyl radical (among other species). The hydroxyl radical is the destructive agent in radiation therapy. However radiation not only kills the cancer cells, but causes extensive damage to normal, healthy tissue as well (**pro-con**). The use of radiation in cancer therapy is indeed a very primitive way to deal with a complex problem.

In the early part of this century Marie Curie identified **chemical carcinogens** in pitch blende. It was subsequently suggested that chemicals may act similiar to X-rays by altering the genetic material, causing mutations.

The available methods for measuring mutations in mammalian cells however were very difficult and time consuming. **Bruce Ames** of the University of California at Berkeley decided to measure the ability of possible mutagens to induce genetic damage in *Salmonella typhimurium*. Detecting mutations in *Salmonella* was far quicker and cheaper than searching for mutations in mammalian genes.

Some time earlier it was established by James and Elisabeth Miller, that many chemical carcinogens are not carcinogens per se, but are converted to the ultimate carcinogenic form by enzymes in our cells. These enzymes are abundant in mammalian liver extracts. First a cell alters the chemical structure of the pro-mutagen to the ultimate mutagen, which then interacts with DNA and alters its information content. Ames therefore added to his test system rat liver extracts to activate pro-carcinogens. This test, known as the **Ames test** (see Varmus and Weinberg, 1993), resulted in a striking correlation between **mutagenicity and carcinogenicity.** Pro-carcinogens can also be converted to ultimate carcinogens non-enzymatically by ROMs.

Asbestos stimulated phagocytes (produce H_2O_2 and $\cdot OH$) can replace the mixed function oxidase from liver extracts in activating benzo(a)pyrene (Roman Franco, 1982). This is why **chronic inflammation** (activated phagocytes) is related to **carcinogenesis**.

Ames tested a huge number of man made and natural compounds for their mutagenicity. He showed that in addition to man made chemicals, like pesticides, there are numerous natural occurring substances, which are mutagenic and carcinogenic (Ames, 1983). This news caused quite a stir among environmentalists. The strategy of producing mutagenic and carcinogenic substances, evolved during millions of years of evolution as defense against predators. It was estimated by Ames that a glass of apple juice contaminated with the chemical Alar (to prevent premature ripening) is far less carcinogenic than a peanut butter sandwich (contains aflatoxin), or mushrooms (contain hydrazine), or a bottle of beer (contains N-nitrosamines, as a natural products of fermentation).

ROMs react with DNA to produce a huge number of products. Whenever DNA gets damaged it may trigger cell death. However, the damage can also be repaired. This repair job is carried out by the **DNA polymerase enzyme**. Whenever DNA polymerase detects a damaged DNA base, it cuts it out and replaces it with the proper base. The damaged part is excreted in the urine, where it may be detected with a variety of analytical techniques. As I have already discussed, the most important marker for oxidative stress is the **8-OHdG**, which is mutagenic. Oxygen radicals damage our DNA some 10^4 times/ day. Most of this damage is repaired. However nothing in this world is perfect, not even the genetic machinery which has been honed to perfection over billions of years of evolution. Some damage does not get repaired and accumulates over a lifetime, leading to aging, cancer and death. Numerous studies have shown that carcinogens injected into rats resulted in an increase in 8-OHdG excretion (Kasai and Nishimura, 1991)

Epidemiology provides a compelling demonstration that human cancers increase with age and that cancer development is a multi-step process. Statistical analysis has shown that the risk of contracting for example colon cancer increases approximately as the fifth power of elapsed time (age). This implies a succession of five distinct events. Similar behavior was observed with many tumors. Cancer is, therefore, one of the degenerative diseases of old age, although exogenous factors may increase (lifestyle) or decrease (low caloric intake, proper nutrition) the cancer incidence. Cancer is, therefore, not the scourge of our modern society, but is the consequence of our better lifestyle (hygiene, nutrition), which keeps us alive longer.

Since **8-OHdG** is formed via ROMs and since ROMs are formed via normal metabolism, there should be a relationship between basal metabolic rate, life span and cancer incidence. As pointed out by Ames, rodents have a higher metabolic rate (higher steady state levels of ROMs), but lower levels of the antioxidant uric acid, compared to humans. A higher metabolic rate would increase the level of endogenous mutagens and could be an important factor in their different longevity. It has been suggested that the higher metabolic rate in rodents (compared to humans) is responsible for the fact that over half of all chemicals tested were found to be carcinogenic in rodents (Ames and Gold, 1990). This means, that a compound shown to be carcinogenic in rodents is not necessarily carcinogenic in humans, who have lower fluxes of ROMs and higher levels of antioxidants. The artificial sweetener saccharin, which was found to be carcinogenic in rodents, has just recently been taken off the carcinogen list by the FDA.

Ames and coworkers (Shigenaga et al. 1989) studied the amounts of **8-OHdG** excreted in the urine of humans, rats and mice. They observed that mice excreted about 3.3 times as much 8-OHdG than humans (582 versus 178 residues/cell/day). This result supports the theory that longevity is related to lower metabolic rate and calorie intake, although this theory has been questioned by some investigators.

The observation of **8-OHdG** in the urine does not tell us where this product comes from (**nuclear or mitochondrial DNA**). DNA damage was assessed in nuclear DNA and in mitochondria exposed to oxidative insult. The amount of 8-OHdG in m-DNA was about 16 times higher than in n-DNA (Richter et al. 1988). This may be due to several factors. The mitochondria consume about 90% of the cells oxygen and produce via the respiratory chain a continuous flux of reactive oxygen species. In addition m-DNA is not protected by a sheath of proteins (histones) and is, therefore, more susceptible to attack. In addition to the increased damage to mitochondrial DNA, it was found that damage to nuclear DNA is repaired more efficiently (Yakes and van Houten, 1997). Why should this be so? The nuclear DNA is obviously more important for the survival of the species than mitochondrial DNA. Without the program nothing works. The results of Yakes and van Houten could mean that damage to m-DNA is the more important factor in aging and age related diseases.

Inflammation, Cell Division and Cancer

Clinicians have recognized for a long time that malignant tumors arose after long periods of **chronic inflammation** (Weitzman and Gordon, 1990). Bowel cancer after ulcerative colitis or Crohn's disease, bladder cancer after schistosomiasis, and gastric cancer after atrophic gastritis. Tumors induced in rodents by a systemically given carcinogen appear preferentially at sites of wounding and inflammation. Tumors induced by the Rous sarcoma virus appears preferentially at the site of injury or inflammation. Inflammation increased tumor formation in the colon, bladder and skin (see Eberhardt, 2000).

Stimulated phagocytes produce abundant amounts of $O_2^{\cdot -}$/ $HO_2^{\cdot}$, H_2O_2, $\cdot OH$, $^1O_2$, $\cdot NO$, NO_2, N_2O_3, and ^-OONO. All of these reactive metabolites can cause damage to important biomolecules including DNA and produce **8-OHdG**, which is highly mutagenic (Cheng et al. 1992). This observation provides us therefore with a rationale why inflammation and carcinogenesis might be related. Human neutrophils and macrophages produce mutations in bacteria and mamallian cells. Neutrophils from patients with **chronic granulomatous disease** (which do not produce reactive oxygen metabolites) (Babior and Crowley, 1983) are not mutagenic, thus clearly implicating these reactive oxygen metabolites as the mutagenic agent.

The involvement of the hydroxyl radical in these DNA base modifications is quite straightforward in solution (*in vitro*). In intact cells (*in vivo*), however, we have a different situation. The question arises: how does the highly reactive hydroxyl radical reach the DNA? The DNA is well protected by a nuclear membrane and a sheath of histones. Among the oxygen metabolites only H_2O_2 can penetrate membranes and the formation of hydroxyl radical must occur in close proximity to DNA involving one-electron transfer from a DNA associated metal ion. I have already discussed this **"site-specific formation"** of hydroxyl radicals (Chapter 2). However, even if H_2O_2 is converted to hydroxyl extracellularly, it still can cause damage to DNA by oxidation of membrane lipids leading to toxic products (aldehydes), which I have already discussed (Chapter 2).

The oxidants produced by **activated phagocytes** are signals for **mitogenesis** (promotion of wound healing). Cell division is critical for mutagenesis and carcinogenesis. During cell division, single stranded DNA is without base pairing or histones and is therefore more sensitive to damage than double stranded DNA. Therefore endogenous or exogenous damage is increased if cells are proliferating. Non-dividing cells, such as adult nerve cells never develop tumors.

If **cell division (mitosis)** is essential for **carcinogenesis**, then agents that can lead to increased cell division should lead to increased tumor formation. Increased cell division can be caused by internal stimuli, like **hormones (estrogen, testosterone)** (Henderson et al. 1988) or external stimuli like drugs, chemicals, bacteria, viruses, such as **hepatitis B and C in liver cancer** or **heliobacter pylori in stomach cancer**, physical and mechanical trauma and chronic inflammation (see Eberhardt, 2000). The controversial role of hormone replacement therapy for postmenopausal women will be discussed in Chapter 6.

The **virus-cancer link** was first observed in the Rous sarcoma virus in 1910. Like all revolutionary ideas it took a long time before the virus-cancer link was accepted by the scientific community. In 1966 **P. Rous** finally received the Nobel Prize for his discovery of tumor-inducing viruses.

Infectious agents and prolonged irritation by chemical and physical agents cause cell death. The subsequent cell division to repair the damaged tissue increases the risk of cancer at the damaged site. Mechanical abrasions of epithelial cells initiates cell proliferation. This type of abrasion has been suggested as causing cancer of the stomach (by rough or salty foods), and gall bladder (stones damage the walls of the gall bladder) (Diehl, 1983). Physical trauma caused by asbestos in lung epithelial cells contributes to increased lung cancer rates. Asbestos may also get caught in the intestinal tract and cause **chronic inflammation**, thus contributing to **bowel cancer.**

Another factor in the etiology of bowel cancer is the formation of **nitrogen oxides** by activated phagocytes

(Chapter 3). Intestinal bacteria produce many different amines, which can react with N_2O_3 to give **carcinogenic N-nitrosamines** (Chapter 2).

Summary. H. Varmus and R. A. Weinberg in their book: "Genes and the biology of cancer", state: the causes of cancer and their links to fundamental aspects of life - the mutagens in our environment and the genes required for normal growth and development - suggests that cancer is intrinsic to multicellular life and that the total eradication of cancer from our species is implausible. A somewhat more pessimistic sentiment was already expressed by Goethe in his *Faust* as stated by Mephisto:

> *the spirit of medicine is easy to know*
> *through the microcosm and macrocosm you breeze,*
> *and in the end you let it go*
> *as God may please.*

Lung diseases

Since the main entry of oxygen into the body is through the lungs, it is not surprising that the relationship between lung and oxidative stress has been extensively investigated. The lung is exposed to bacteria, viruses, inorganic particles (asbestos, silica), toxins like paraquat, CCl_4, bleomycin, anthracyclin and toxins which are part of air pollution, like the nitrogen oxides (NO_2, N_2O_3), ozone (O_3) and airborne particulates, such as benzo(a)pyrene (B(a)P). Even if certain pollutants in air are present in low concentrations, due to the huge volume of air passing through the lung and the large surface area, the lung is undergoing considerable oxidative stress.

As I have pointed out before the antioxidant defenses are not evenly distributed throughout the body The antioxidants are located where they are needed most. In defense against oxidative stress the lung uses all of the available defenses, as discussed in Chapter 4 (see Eberhardt, 2000).

Erythrocytes are not only carriers of oxygen, but also fulfill important defensive function during oxidative stress. In conditions of high oxidative stress (when ROMs overwhelm the antioxidant defenses of pulmonary resident cells, the lung can recruit additional antioxidant defenses from circulating blood cells (**erythrocytes, platelets and phagocytes**). Erythrocytes are rich in SOD, CAT and GSH-Px. Erythrocytes also have anion channels, which permit the entry of $O_2^{\cdot-}$ (an anion) into the cells, where it can be detoxified via SOD and CAT to H_2O, thus preventing its conversion to the more toxic $\cdot OH$ and HOCl. The first SOD was discovered in erythrocytes by McCord and Fridovich (1969).

I have already discussed that $O_2^{\cdot-}$ induces the formation of antioxidant defenses. This makes good economic sense. If no ROMs are produced, a cell does not have to waste its energy and resources to manufacture antioxidants (SOD, CAT, GSH-Px). A nation without enemies does not have to invest billions in defense.

Reactive oxygen metabolites induce in erythrocytes the formation of more GSH and CAT. It was observed that erythrocytes of smokers contained more GSH and CAT than those of non-smokers (Toth et al., 1986). This example shows that the lungs are not helpless against the constant onslaught by ROMs, and that an occasional cigar won't kill us!

However erythrocytes can also have a **prooxidant effect**. ROMs cause lysis of erythrocytes, thus releasing not only their antioxidant enzymes into the microenvironment, but also some hemoglobin-bound iron, which can serve as a catalyst for hydroxyl radical ($\cdot OH$) formation.

Platelets are also rich in CAT and have several times more GSH than erythrocytes. Hydrogen peroxide causes aggregation of platelets and may serve therefore as a signal to recruit platelets to the site of high oxidative stress. There is also evidence for a protective role of platelets *in vivo*. It has been shown that in rats lung toxicity induced by α-naphthyl-thiourea is potentiated in absence of circulating platelets.

Platelets also synthesize nitric oxide ($\cdot$NO), which depending on the special circumstances can react via its metabolites (NO_2, N_2O_3, ^-OONO) as an oxidizing agent or as a radical scavenger. The radical scavenging ability may contribute to the antioxidant effect of platelets.

In addition to erythrocytes and platelets it has been suggested that **phagocytes** contribute to pulmonary **antioxidant defense**. The possibility that phagocytes, the most abundant source of ROMs (Chapter 3) can act as an antioxidant seems strange indeed, but truth sometimes is stranger than fiction. Phagocytes contain considerable amounts of antioxidants in order to protect themselves against their own ROMs. Phagocytes have indeed been shown to decrease H_2O_2 *in vitro*. Phagocytes attracted to areas of high oxidative stress (chemotaxis) may lyse and loose their ability to form ROMs while releasing their antioxidants into the microenvironment. We have a **pro and con** situation (see Eberhardt, 2000).

The lungs are the main port of entry for bacteria and viruses. Bacteria and viruses are attacked by phagocytes, which produce copious amounts of ROMs (Chapter 3). While in bacterial infection the damage inflicted on the surrounding tissue is minimal it is quite substantial in viral infection. Viruses leave many battle scars, which affect the viability of the lung. On the other hand inorganic particles, like asbestos fibers or silica dust undergo failed phagocytosis. In these cases the particles cannot be destroyed and the phagocytes release ROMs into the microenvironment, providing a continuous source of damaging ROMs. Asbestos fibers can be considered acting like an implanted radiation source, producing hydroxyl radicals over the lifetime of the organism.

Since oxygen is metabolized to a number of highly reactive metabolites, it is not surprising that **hyperbaric oxygen** causes tissue damage and alveolar dysfunction. The damaging effect of hyperbaric oxygen was noted long before the discovery of superoxide and superoxide dismutase. **Gershman et al. (1954)** established a connection between hyperbaric oxygen damage and X-irradiation. They proposed that the damage is due to increased formation of oxygen radicals at a rate in excess of the antioxidant capabilities.

With increased oxygen tension the oxygen consumption increases, and the fraction of oxygen converted to ROMs is also increasing. Antioxidant depletion decreases survival of animals exposed to hyperoxia. However, **sublethal hyperoxia** induces antioxidant defenses and increases survival. Remember: small amounts of O_2^- induces formation of antioxidants (GSH) as discussed in Chapter 4 under SOD paradox.

The primary target of hyperoxia-generated ROMs are lipids. This results in increased membrane permeability, affecting cellular **homeostasis**. However the damage caused by hyperoxia can also be put to good use. In **radiation therapy** increased oxygen tension has been extensively used to increase the killing of tumor cells.

The opposite effect (decrease in oxygen supply) has recently been used to kill cancer cells. Oxygen provides us with the necessary energy for all metabolic processes, and is thus essential for life. In order for a tumor to grow it needs energy (more energy than a normal cell). This is accomplished by building a network of new blood vessels, a process called **angiogenesis**. Some drugs, which inhibit angiogenesis are Vioxx and Celebrex, which have already been used for some time as treatments for arthritis. Thalidomide is another potent angiogenesis inhibitor, which is already widely used against multiple myeloma.

There is no doubt that **smoking** is bad for our health. Smoking increases the risk of coronary heart disease, emphysema, asthma and lung cancer Cigarette smoke contains reactive oxygen species, as well as many toxic substances, like aldehydes, epoxides, endoperoxides, B(a)P, hydrogen peroxide and nitrogen oxides, which lead to an increase in lipid peroxidation (as was found in plasma of smokers) and to an increase in urinary excretion of **8-OHdG**. Nakayama et al. (1985) showed that cigarette smoke induces a considerable number of **DNA single strand breaks** in cultured human cells. I have already discussed the formation of 8-OHdG via ROM-producing systems (Chapter 2). Cigarette smoke was found to considerably increase the yield of 8-OHdG (this molecule is mutagenic).

Vitamin E supplementation was found to decrease lipid peroxidation This protective effect against lipid peroxidation however does not translate into a decreased risk for lung cancer in smokers. In a Finnish study on male smokers Vitamin E did not prevent lung cancer and Vitamin A may even be harmful (Heinonen et al., 1994).

When small **asbestos fibers** are inhaled, the immune system reacts to deal with the foreign intruder. Macrophages try to digest the particles, but because asbestos is an inorganic fiber this cannot be accomplished. Instead we have a process known as **"failed phagocytosis"**. The phagocytes produce ROMs, which are released into the microenvironment. The asbestos fibers react with H_2O_2 to give $\cdot OH$ (Eberhardt et al. 1985). The reaction involves Fe- or other metal ion sites on the asbestos surface.

Asbestos fibers lodged in the airways or in the intestine provide a continuous source of radicals (like an implanted radiation source), and contribute to lung and intestinal cancers.

Asthma is a common disease in young people under 20 years of age. Statistics show a considerable increase in diagnosed asthma cases during the last decades. Like the previously discussed diseases of the lung, asthma is caused by external factors, like smoking, second hand smoking and air pollution.

Low dietary intake of Vitamin C has been consistently associated with increased asthmatic symptoms. As pointed out before the amelioration of symptoms by antioxidants is an indication that ROMs are involved in the disease process. The Vitamin C content of plasma and blood leukocytes was examined in asthmatic and normal patients. The asthmatic patients had 35% less vitamin C in their blood leukocytes and 50% less in their plasma (Olusi et al., 1979). The low Vitamin C levels are not the cause of asthma, but a consequence of overexposure of the lung tissue to a variety of ROMs and toxins, which react with Vitamin C, thus reducing its concentration.

Environmental pollutants can react directly with vital targets, but they also can stimulate leukocytes to produce ROMs. It was proposed that asthma involves an **overproduction of ROMs by leukocytes**. Eosinophils, alveolar macrophages and neutrophils from asthmatic patients produce more ROMs compared to normal patients (Barnes, 1990). Epithelial cells of course produce nitric oxide, a vasorelaxant. The importance of activated leukocytes in asthma is also evident from the observation that asthmatic patients exhale increased amounts of nitric oxide, a product of activated leukocytes (Chapter 3).

Ischemia-Reperfusion

Ischemia (lack of oxygen) causes tissue necrosis in many organs, such as heart, intestine, brain, kidney and lung. Permanent deprivation of oxygen supply (blood flow) is lethal to any tissue or organ. Any interruption of the oxygen supply should therefore be immediately treated with reperfusion. This is especially true for the brain, which consumes about 18% of the total oxygen. Here again we have another **pro and con** situation. Reoxygenation is necessary, but causes additional damage. This observation has become known as the "**oxygen paradox**". Reperfusion causes additional tissue injury due to formation of reactive oxygen metabolites. Increased amounts of pentane (a biomarker for lipid peroxidation) was found in the breath of patients, who suffered a stroke during the preceeding 12 hours (Weitz et al. 1991).

There are many different causes for ischemia. It can result from **atherosclerosis** (narrowing of the arteries), thrombo-embolism or external pressure on blood vessels (as in the case of tumors). Ischemia can also be **iatrogenic**. It can occur during surgery, when the blood flow to an organ must be interrupted. Ischemia is a lack of oxygen and substrate and therefore a lack of aerobic energy production. The ATP (energy storing molecule) content of the tissue falls rapidly. Since ATP is essential for vital metabolic processes, a decrease in ATP initiates a cascade of damaging effects. The ischemic tissue attracts **neutrophils (chemotaxis)**, which cause further blockage of capillary arteries. The accumulating neutrophils upon reperfusion cause additional damage via the formation of **oxygen radicals** ($O_2^{\cdot-}$ and $\cdot OH$) singlet oxygen, HOCl, NO and ^-OONO. Ischemia causes large increases in the permeability of the endothelium to macromolecules.

Ischemia-reperfusion (I/R) injury may occur in any organ of the human body. From a medical point of view, however, the most significant organs are **the heart and the brain**. Heart attacks and strokes are among the leading causes of death in the USA and other Western countries. Oxygen radicals are definitely not a major contributor to cell killing during ischemia, because oxygen is unavailable in ischemia. A dramatic decrease in intracellular ATP production correlates best with irreversible injury. Ischemia-reperfusion (I/R) represent two sides of the same coin. Reoxygenation damage depends on previous ischemia.

Since ·OH is so highly reactive there are innumerable compounds, which can scavenge ·OH and thus attenuate tissue damage in I/R. However for practical purposes, we are limited to compounds, which are non-toxic and can therefore be added in sufficient concentration to cardioplegic solutions during reperfusion. Some of the compounds, which have been examined are **α-tocopherol, mannitol, allopurinol, oxypurinol and DMSO.**

Ever since the discovery of $O_2{\cdot}^-$ and SOD by McCord and Fridovich (1968 and 1969) has SOD been used extensively for the identification of $O_2{\cdot}^-$ *in vivo*. In myocardial I/R injury, however, the cardioprotective effect of MnSOD was lost at high doses in the reoxygenated heart. McCord and coworkers (1995) recognized the dual character of the superoxide radical anion, which can act as a **chain initiator** or a **chain terminator** in lipid peroxidation (see Chapter 4, SOD paradox). This realization leads to an optimum SOD concentration. Too little SOD or too much SOD will have serious consequences.

Leukocytes play an important role in I/R injury. Leukocytes accumulate in tissues exposed to I/R. polymorphonuclear leukocytes (PMNs) provide the most abundant source of ROMs. It is therefore not surprising that the accumulation of leukocytes in ischemic tissue leads to the generation of ROMs upon reperfusion, and subsequent tissue damage (see Eberhardt, 2000).

The role of iron in I/R injury

Since the formation of hydroxyl radicals is accelerated by Fe(II) or Cu(I), the control of these metal ions plays an important role in many diseases, including ischemia-reperfusion injury. During cardiac surgery most animal and clinical studies have demonstrated the role of deferoxamine (an iron chelator) in reducing generation of ROMs . However, this effect has not been translated into improved postoperative patient outcome. There is considerable evidence showing that **iron overload** is a risk factor in many diseases, such as **colorectal** and **lung cancer**, **heart disease** and **Parkinson's disease**. I have already discussed the damaging effects of Fe^{2+} and Cu^+ in Chapter 2.

Neurodegenerative Disorders

General considerations

The CNS is particularly vulnerable to oxidative stress for several reasons. The brain consumes about 18% of total oxygen, but represents only about 2% of the total body weight. This indicates increased oxygen metabolism (higher ROM levels), but at the same time the CNS is low in antioxidants (SOD, CAT, GSH-Px) and high in unsaturated fatty acids.

Interruption of the oxygen supply (trauma, atherosclerosis, stroke) has serious consequences. It was observed that hypoxia (low oxygen tension) caused the release of dopamine from dopaminergic neurons. **Dopamine** is oxidized by the enzyme **monoamine peroxidase** (MAO) to give H_2O_2. The formation of H_2O_2 from dopamine is an important step in **Parkinson's disease**. Specific areas of the human brain (substantia nigra, globus pallidus) are rich in iron. Therefore, any injury to the brain by trauma or stroke will release iron ions and lead to formation of **hydroxyl radicals** (Chapter 2).

In addition, the CNS contains microglia, which are the ontogenetic and functional equivalents of mononuclear phagocytes in somatic tissue. **Microglia** are an abundant source of ROMs in the central nervous system. Appropriately stimulated microglia produce superoxide as well as ·NO and cause cell injury. In stimulated microglia we have the right conditions for peroxynitrite formation (high concentrations of O_2·⁻ and ·NO), which is the most destructive metabolite of ·NO. **Peroxynitrite** reacts with any molecule it encounters (lipids, proteins and DNA).

Increased microglia activity has indeed been associated with several neurodegenerative diseases. These diseases are excellent examples of the disturbance of the **prooxidant/antioxidant balance**. The over- or under-producion of prooxidant and antioxidant enzymes is well documented in **amyotrophic lateral sclerosis (Lou Gehrig's disease), Alzheimer's disease, Parkinson's disease, Down's syndrome and multiple sclerosis**. In these diseases we find examples of low SOD (ALS, AD), high SOD (PD, DS) and high NOS (MS).

Amyotrophic lateral sclerosis (ALS)

The disease, popularly known as Lou Gehrig's disease is a highly debilitating disease with lethal consequences. It is characterized by generation of motor neurons with progressive paralysis.

ALS is caused by a genetic defect, which causes an underproduction of one of the superoxide dismutase enzymes (the CuZnSOD). The lower SOD levels lead to increased levels of reactive oxygen metabolites (ROMs). We can envisage the following sequence of events:

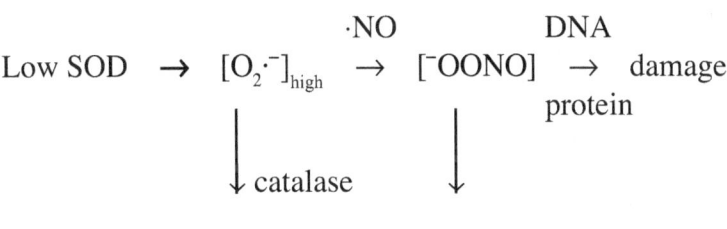

Other evidence has shown that the disease is caused by the production of an altered protein (caused by the-altered gene), which is toxic to the neurons. For a more detailed discussion and references see Eberhardt (2000).

Alzheimer's disease

Alzheimer's disease is an age-related disease affecting the central nervous system. It has been estimated that if the average life expectancy reaches 100 about one-third of the population will be affected by Alzheimer's disease (AD). In addition to aging AD has a genetic factor.

Lipofuscin and ceroid accumulate in liver, heart and brain as aging progresses and these pigments also increase in Alzheimer's dementia In AD patients we observe low levels of CuZnSOD and high steady state concentrations of $O_2^{\cdot-}$. High concentrations of $O_2^{\cdot-}$ favor combination with $\cdot$NO to form the highly destructive $^-$OONO. Since in AD patients the **antioxidant defenses** (SOD, CAT, GSH-Px, GSH, Vitamins C and E, uric acid, selenium) are low the reactive species $^-$OONO and $\cdot$OH are not scavenged, but react with lipids, proteins, and DNA and cause extensive damage. The formation of peroxynitrite in AD is evident from the formation of 3-nitrotyrosine (Chapter 4) in the neurofibrillatory tangles in the brains of Alzheimer's patients (Good et al., 1996). An **altered DNA** causes the production of an **altered protein**. One such protein is the **β-amyloid protein in Alzheimer's disease**. Several observations led to the hypothesis that oxidative stress is involved in the toxicity of β-amyloids. After exposure of cells to β-amyloids, an increase in H_2O_2 was detected and Vitamin E and CAT protected the cells (Behl et al., 1992, 1994). The amyloid stimulates **nitric oxide synthase** (NOS), the enzyme responsible for the synthesis of nitric oxide ($\cdot$NO). The amyloid protein can also break apart to yield radical fragments, which in turn cause cellular damage. Recently it was found (Huang et al. 1999), that the amyloid is a complexing agent for Fe(III) and Cu(II) ions. **Cu(II)** is reduced by the amyloid to **Cu(I)**, which in turn reacts with oxygen to yield hydrogen peroxide (H_2O_2) (Eberhardt, 1991). It has been known for some time that Cu(I) reacts with H_2O_2 in a Fenton-type reaction to give the highly reactive hydroxyl radical (Eberhardt et al., 1989).

Products of **tryptophan catabolism** (the 3-hydroxy-kynurenine, 3-hydroxyanthranilic acid and quinolinic acid) have been demonstrated to reduce Cu(II) and Fe(III) to Cu(I) and Fe(II) resp. We have therefore the possibility that the catabolites of tryptophan act as a cofactor in oxidative damage to proteins, such as **α-crystallin** (present in the lens of the eye) and thus initiates formation of cataracts (Goldstein et al. 2000). The intermediacy of tryptophan catabolites may also be involved in a variety of neurodegenerative disorders, such as Huntington's disease, Parkinson's disease, and multiple sclerosis among others. It is interesting to point out that diseases with such different clinical manifestations (**cataract formation, PD, MS, AD**) have the same common chemical mechanism (reduction of Cu^{2+} to Cu^+, formation of H_2O_2 and formation of ·OH).

Multiple sclerosis

Chronic inflammation is a major factor in disease progression not only in the CNS, but also in other organs like the gastrointestinal tract, liver and lung. As I discussed in Chapter 5, phagocytes are the most abundant source of $O_2^{·-}$, H_2O_2, $^1O_2$, ·OH, HOCl, ·NO, and ^-OONO. The phagocytes are stimulated by bacteria, viruses, endotoxins and cytokines. Therefore, whenever we have inflammation we have a change in the prooxidant/antioxidant balance with damaging consequences.

Chronic inflammation is involved in bowel cancer as well as in multiple sclerosis. **Microglial cells**, which are the macrophages of the CNS are activated in MS lesions (Bö et al., 1994). Activated microglial cells produce more NO. Metabolic balance studies have shown excess formation of NO_2^- and NO_3^- in patients with fever long before the importance of ·NO in inflammation was recognized

In MS we have an **overproduction of NOS** (nitric oxide synthase) producing high levels of NO and peroxynitrite. A knowledge of the mechanism of ROM formation and destruction shows us how we can possibly intervene therapeutically at different stages of disease progression. We can inhibit NOS (NOS inhibitors like L-arginine analogs), or we can scavenge the peroxynitrite ($^-$OONO). Amino-guanidine, an inhibitor of inducible NOS was indeed found to ameleriorate experimental autoimmmune encephalomyelitis (EAE) in mice. EAE is the animal model for MS.

In MS the symptoms are **demyelination** of the myelin sheath of neurons. This demyelination is caused by $^-$OONO. The presence of **3-nitrotyrosine** residues, a biological marker for $^-$OONO, as discussed in Chapter 3, was observed in higher concentration in brain tissues of MS patients (Bagasra et al., 1998). We should, therefore, be able to ameliorate MS symptoms by scavenging $^-$OONO. I have already discussed the role of uric acid as such a scavenger (Chapter 5). Patients with MS had significant lower levels of serum uric acid than controls. Statistical evaluation of over 20 million patient records for the incidence of **MS and gout (hyperuricemia)** revealed that the two diseases are almost mutually exclusive, supporting the peroxynitrite scavenging by uric acid (Hooper et al., 1998). The excess formation of $^-$OONO also implies increased lipid peroxidation (Chapter 2) as evidenced by increased exhalation of **ethane and pentane** by MS patients (Toshniwal and Zarling, 1992).

As I have discussed in Chapter 2, the formation of the highly destructive hydroxyl radical requires **transition metal ions**, such as Fe^{2+} or Cu^+. In this connection it is interesting to note, that an increased deposition of hemosiderin (an iron-complexing protein) in brain of MS patients was observed (Adams, 1988). Hemosiderin is believed to be a large aggregate of ferritin molecules with a much higher content of iron. The involvement of homosiderin in the formation of hydroxyl radicals is another possibility for oxidative stress.

Parkinson's disease

Dopamine, a metabolite of the amino acid tyrosine, is a dihydroxy-aromatic compound, and a neurotransmitter. It has long been recognized as a source of ROM's. Dopamine like other polyhydroxy-aromatics is easily autoxidized non-enzymatically to yield superoxide radical anion.

It has been observed that **hypoxia** (low oxygen pressure) induces the release of dopamine from the storage vesicles and thus may cause PD. Dopamine is metabolized in the mito-chondria by the enzyme **monoamine oxidase** to yield H_2O_2 (Knight, 1997). The H_2O_2 formed in this reaction may react with Fe^{2+} or Cu^+ or with CAT:

$$H_2O \quad \xleftarrow{\text{CAT}} \quad H_2O_2 \quad \xrightarrow{Fe^{2+} \text{ or } Cu^+} \quad \cdot OH$$

We have a competition between CAT and metal ions for H_2O_2. Since in Parkinson's disease CAT and GSH-Px levels are low and iron levels are high, the winner in this competition is obvious.

Parkinson's disease is characterized by the following abnormalities: An **increase in CuZnSOD**, which is prefer-entially expressed in the neuromelanin containing neurons of substantia nigra, leads to **low $[O_2^{·-}]_{ss}$** (Ceballos et al., 1990). This subset of neurons is known to be particularly vulnerable to degeneration in PD. The observation that an increase in SOD (an antioxidant) and a decrease in $[O_2^{·-}]$ (a prooxidant) causes oxidative damage has become known as the **SOD paradox**. SOD also fulfills a regulatory function and low $[O_2^{·-}]$ leads to lower antioxidant defenses like GSH, which was indeed observed in PD patients. GSH in the substantia nigra was lower by 40% compared to controls (lower GSH/GSSG ratios). I have already discussed the **neurotoxicity of GSSG** (Chapter 2)

Since the ·OH radical is highly reactive and reacts in close proximity of its site of formation (site-specific), the damage occurs preferentially in these neuromelanin-containing neurons of substantia nigra.

Atherosclerosis

Atherosclerosis is the main cause of death in the USA and Western Europe. Narrowing of the arteries by atherosclerotic plaques causes myocardial infarction or stroke. Atheros-clerosis is a progressive disease characterized by the accumulation of lipids and fibrous elements in the large arteries. The early lesions of atherosclerosis consist of subendothelial accumulation of cholesterol-filled macrophages, known as **foam cells**. Since macrophages are the most prolific producers of ROMs (Chapter 3), the presence of choleste rol-filled macrophages (foam cells) in the intima immediately raises the possibility of ROM involvement in atherogenesis (Lusis, 2000).

Cholesterol as a major constituent of membranes is vital for cell growth and survival. However excessive amounts of cholesterol can be lethal, since it is intimately associated with the development of atherosclerosis. Because cholesterol is essential for cell growth, it is synthesized by our own cells. This synthesis is very complex involving 25 different reaction steps, using acetate (vinegar) as the building block. We not only take in cholesterol with our food, but it is synthesized by our own metabolism.

Full blown atherosclerosis can be produced in experimental animals simply by feeding it a diet rich in cholesterol. The first observable change in the artery following the feeding of a high-fat, high-cholesterol diet is the accumulation of **low density lipoprotein particles (LDL)** in the intima. LDL consists of 75% lipids and 25% protein. LDL is the most important **transport protein of cholesterol**. It provides the cells with the cholesterol needed for the synthesis of membranes and steroid hormones.

Within days monocytes attach to the surface of the endo-thelium. The monocytes then migrate across the endothelium into the intima, where they proliferate, differentiate into macrophages and take up the lipoproteins leading to foam cell formation. A schematic representation of an arterial wall is shown in Fig. 3. The wall of an artery consists of three main layers (from the inside to the outside): the intima, media and adventitia. The intima is covered with a monolayer of endo-thelial cells (EC). The intima consists of of connective tissue (collagen and protoglycans). The media consists of smooth muscle cells (SMCs), and the outermost layer, the adventitia, consists of connective tissue interspersed with fibroblasts and smooth muscle cells (Lusis, 2000).

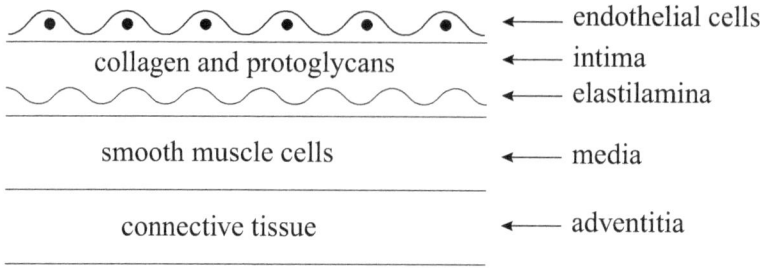

Fig. 3. Arterial cell wall

More advanced lesions are characterized by the accumulation of lipid-rich necrotic debris and smooth muscle cells (SMCs). Advanced lesions can grow sufficiently large to block blood flow, however the most frequent clinical complication is an acute occlusion due to the formation of a thrombus or blood clot, resulting in myocardial infarction or stroke.

In vivo tissue macrophages take up large amounts of LDL cholesterol (cholesterol attached to the transport vehicle, the LDL via an ester bond) and are transformed into foam cells. On the other hand *in vitro* (in a Petri dish) tissue macrophages take up native LDL at extremely slow rates This observation has become known as the **LDL-macrophage paradox**. Paradoxes arise as the result of incomplete knowledge. The resolution of paradoxes leads to a deeper understanding of the processes involved.

Before LDL can be taken up by macrophages it has to be oxidatively modified. The key event in atherogenesis is **oxidative modification** of LDL and endocytosis by macrophages to form foam cells (Steinbrecher et al., 1990). This modification involves reactive oxygen metabolites (ROMs) produced by endothelial cells and macrophages. In addition several enzymes (such as myeloperoxidase) are involved in oxidative modification. MPO produces hypochlorous acid (HOCl) (Chapter 3).

The major targets of ROMs in biological systems are lipids, proteins and DNA. In the case of atherogenesis the initiating event is attack on the low density lipoprotein (LDL).

The two primary oxygen metabolites $O_2^{\cdot-}$ and H_2O_2 are not very reactive. The conversion to the more reactive secondary metabolites requires **transition metal ions** as catalysts. The metal ions are complexed to a variety of proteins. In addition the metal ions are usually present in their higher oxidation states and have to be converted to their lower oxidation state by reducing agents ($O_2^{\cdot-}$, AH_2, TH, RSH, flavonoids). The importance of transition metal ions in the modification of LDL by cultured cells is clearly established. PUFAs are easily autoxidized in an oxygen atmosphere and accumulate lipid hydroperoxides (rancification).These peroxides in presence of Cu^{2+}/Cu^+ initiate further lipid peroxidation and provide an explanation for the observation by Salonen et al. (1991), that high plasma Cu^{2+} levels increase the risk of CAD. In this case **Vitamin E** may act as a **prooxidant**, reducing Cu^{2+} to Cu^+ with subsequent formation of hydroxyl radicals (Chapter 3).

LOOH in presence of Cu^{2+} can initiate lipid peroxidation:

$$LOOH + Cu^{2+} \rightarrow LOO\cdot + H^+ + Cu^+$$

The reduction of Cu^{2+} to Cu^+ can also be accomplished by Vitamin C (AH_2), Vitamin E (α-TOH) or $O_2^{\cdot-}$.

Atherogenesis is another textbook example of the disturbance of the prooxidant/antioxidant balance. Increased formation of prooxidants ($\cdot NO$, ^-OONO, $O_2^{\cdot-}/HO_2\cdot$, H_2O_2, $\cdot OH$, $^1O_2$, and HOCl) by a variety of cell types, such as endothelial cells (EC), smooth muscle cells (SMCs), monocytes (MOC) and macrophages (MPH) leads to increased oxidative modification of LDL. Ox-LDL is internalized by macrophages (located in the subendothelial space), whch are converted to foam cells.

Another way to increase oxidative stress is a decrease in anti-oxidants. Important naturally present antioxidants are SH compounds, like cysteine and glutathione. GSH decreases the prooxidant LOOH, but also decreases the GSH/GSSG ratio:

$$2\,GSH\ + LOOH\ \rightarrow\ GSSG\ +\ LOH\ + H_2O$$

The effect of a low GSH/GSSG ratio has already been discussed (Chapter 2). We should be able to counteract an increase in $[prooxidant]_{ss}$ with an increase in $[antioxidant]_{ss}$ by dietary supplementation (Vitamins A, C, E). However, *in vivo,* both **Vitamin C and E** may represent a double-edged sword, since these vitamins may act under certain conditions as **prooxidants** as well (Chapter 4).

Large scale epidemiological studies (39 910 men over a period of four years) (Rimm et al., 1993) have shown a considerable reduction in the clinical expression of coronary artery disease (CAD) and atherosclerosis in patients treated with lipid soluble dietary antioxidants. Intake of **Vitamin E** was associated with lower risk of CAD. The maximal reduction in risk was seen in men consuming 100-249 IU per day, with no further reduction at higher doses (more is not always better!). In this study carotene intake was inversely associated to risk only in smokers and high intake of **Vitamin C** had no effect on CAD. In a similiar study in women Vitamin E lowered the risk of major CAD by 41% (Stampfer et al., 1993). In conclusion these epidemiological studies in humans support the hypothesis that the lipid-soluble antioxidant **Vitamin E** reduces the clinical expression of atherosclerosis and that ROMs must be involved in its progression.

Vitamin C preserves the endogenous antioxidants and quenches oxidants in hydrophilic environments. The oxidation of LDL particles, however, occurs in the subendothelial space (the intima), a hydrophobic environment that favors a protective effect of fat-soluble vitamins (Vitamin E and carotene) over the water soluble Vitamin C.

In addition to Vitamin E there are other protective mechanisms involved in LDL oxidation. **High density lipoprotein (HDL)** has been shown to protect LDL. Epidemiology established an inverse correlation between plasma HDL levels and the risk of coronary heart disease (Miller, 1980). On the other hand elevated levels of plasma LDL correlated with increased risk of coronary heart disease. The HDL cholesterol has therefore become popularly known as the **"good cholesterol"** and LDL-cholesterol as the **"bad cholesterol"**. As I shall discuss under pros and cons of exercise (p.152), moderate exercise combined with diet (low fat, low cholesterol and Vitamin E supplementation) raises the HDL/LDL ratio, thus lowering the risk for coronary artery disease.

Trans fatty acids increase LDL and decrease HDL-cholesterol. Introduction of trans fatty acids into the diet by processed foods was responsible for the dramatic increase in coronary artery disease (Willett et al. 1993, Hu et al. 1997). The damaging effect of *trans* fatty acids is not too surprising. We have evolved over millions of years without *trans* fatty acids, and our metabolism has no use for them. The connection between evolution and nutrition has been discussed in detail by Crawford and Marsh (1989).

The formation of ox-LDL in the vascular wall leads to impaired **EDRF** (endothelium-derived relaxing factor) action. It has been well documented that ·NO released from endothelium by certain stimuli causes vasorelaxation and inhibits platelet aggregation (Chapter 2). Endothelium derived relaxation is impaired in animals with atherosclerosis and in isolated atherosclerotic human arteries (Bossaller et al., 1987).

Inhibition of the vasorelaxant effect means increased platelet aggregation and coronary artery disease.

An interesting consequence of the damaged EDRF response in atherosclerosis is the observation that mental stress induces vasoconstriction (Young, A. C. et al, 1991). Several mechanisms leading to impaired EDRF action are discussed by Eberhardt (2000).

Since **·NO is a radical** it can combine with another radical, i. e. it can act as a **radical scavenger**. It can scavenge other radicals, especially the lipid peroxyl radicals, which are the chain carriers in lipid peroxidation (Chapter 2):

$$LOO· \; + \; LH \; \rightarrow \; LOOH \; + \; L· \quad \text{propagation}$$

$$LOO· \; + \; ·NO \; \rightarrow \; LOONO \quad\quad \text{termination}$$

Nitric oxide can therefore protect LDL from oxidative modification. A diet supplemented with **L-arginine** (the precursor of ·NO) has indeed been shown to lower the risk of atherosclerosis in hypercholesterolemic rabbits (Cooke et al., 1992). On the other hand ·NO deactivation products like nitrogen dioxide or peroxynitrite (NO_2 or ^-OONO) can initiate lipid peroxidation (Chapter 2). The answer to the question: is NO good or bad, depends on the specific circumstances.

As pointed out before the study of the abnormal helps us understand the normal. Studies of **familial hypercholestero-lemia** helped unravel the pathways that regulate plasma cholesterol metabolism, knowledge of which was important in the development of cholesterol lowering drugs. **Tangier disease**, a rare recessive disorder, is characterized by the complete absence of HDL (Lusis, 2000). This disorder shows the importance of HDL as a protective agent. The beneficial effect of nitric oxide in atherogenesis is evident from results in

mice lacking NO synthase. These mice show enhanced atherosclerosis, due in part, to increased blood pressure (Lusis, 2000).

Summary.

The best prescription for prevention of atherosclerosis is: low fat intake, low cholesterol intake and sufficient (whatever that means) intake of Vitamin C, E and selenium. Remember: too little or too much Vitamin C and E can be damaging. However, as pointed out by Steinberg et al. (1989) no matter how successful we deal with hypercholesterolemia (high levels of cholesterol), coronary artery disease will not disappear, because a high cholesterol level is by no means the only causative factor. The importance of genetics and environment in human CAD has been examined in many family and twin studies. Within a population, the heritability of atherosclerosis (the fraction of disease explained by genetics) has been high in most studies, frequently exceeding 50%. Population migration studies, on the other hand, clearly show that the environment (lifestyle) explains much of the variation in disease incidence between populations. CAD results most likely from a **combination of environment, lifestyle, genetics and our increased lifespan**.

Testing for LDL, HDL and blood pressure has long been advocated as a way of identifying individuals at increased risk. However, given the importance of environment and lifestyles (for 246 risk factors see Hopkins and Williams 1981) and the complex genetic etiology of atherosclerosis, the eradication of this disease does not appear likely in the foreseeable future.

OXIDATIVE STRESS AND LIFESTYLES

Everything is poison, it just depends on the dose.
Paracelsus

A change in the prooxidant/antioxidant balance can be brought about by genetic abnormalities or by external factors such as bacteria, viruses, toxins, or air pollution. While there are many factors, which we cannot control, there are some factors which we can control. I like to discuss now how **exercise, diet, alcohol consumption** and **psychological stress** affect the prooxidant/antioxidant balance. These lifestyle factors are excellent examples of the importance of balance in everything we do.

Exercise

No hay nada que dure mas, que un vago bien cuidado
Puerto Rican saying
Nothing lasts longer than a lazy man
who takes care of himself.

Exercise initiates a complex physiological and biochemical response. In our exercise conscious society we have to ask the important question: is exercise good or bad for our health? In my opinion nobody can give an unequivocal answer. The answer depends on many factors. We are dealing with a complex dynamic system where small changes in initial conditions can have unforeseen consequences. The oxidative stress imposed on a tissue depends on the tissue involved, on the age of the individual, on his or her overall health and training status, on the type of exercise, on its intensity and duration (Møller et al., 1996).

Exercise requires **energy**, which is obtained through the increased flow of electrons in the **mitochondrial respiratory chain** leading to increased formation of $O_2{\cdot}^-$ and H_2O_2 and increased formation of **ATP** (the battery of the cell). Continuous exercise then depletes the oxygen supply and the ATP pool, leading to accumulation of ADP.

Oxidative stress results in damage to **lipids, proteins** and **DNA**. The biological markers for lipid peroxidation are **malonaldehyde** or **ethane** and **pentane** in the exhaled breath (Chapter 2). DNA damage is assessed by measuring **8-OHdG** (Chapter 2). Increased formation of ethane and pentane in the exhaled breath in humans undergoing strenuous exercise was indeed observed (Dillard et al 1978). The amount of ethane and pentane was decreased in presence of Vitamin E.

With a single bout of exercise it appears that moderate exercise does not induce oxidative stress, but high intensity exercise does (Everything is poison, it just depends on the dose). Oxidative stress appears to be related to exercise intensity. High intensity training induces muscle damage (eccentric more than concentric). Marathon running increased excretion of 8-OHdG (Alesio, 1993), but moderate cycling exercise did not show any effect (Viguie et al., 1993).

In addition to 8-OHdG (from DNA damage) and ethane and pentane (from lipid peroxidation), there are increased serum levels of some enzymes, which are indicative of oxidative stress. These enzymes are **creatine kinase (CK)** and **lactate dehydrogenase (LDH)** (Kanter et al., 1988) How do these enzymes arise from oxidative stress? Exercise leads to oxygen depletion and accumulation of ADP. At low levels of oxygen, the cells cannot obtain the necessary energy through oxidative phosphorylation (the mitochondrial respiratory chain), but have to switch to another process of energy production. In this process skeletal muscle uses glucose as an energy source and breaks it down to lactic acid (**glycolosis**). Lactate dehydrogenase (LDH) catalyzes the last step in glycolysis, forming lactate from pyruvate. The observation of increased levels of

LDH is therefore an indication that the cell has switched to an **anaerobic metabolism.**

The **creatine kinase** catalyzes the formation of **ATP from ADP** (it recharges the battery). The increase in these enzymes in the serum indicates membrane damage, which is caused by lipid peroxidation. In long distance runners (80 km) an increase in CK and LDH in their blood plasma has indeed been observed.

Another source for increased ROM-induced damage during exercise is the availability of **free iron ions**. Homeostatic control of iron and copper ions plays an important part in ROM induced damage during exercise. Increased formation of $O_2^{\cdot-}$ causes release of free Fe-ions from iron-storage proteins (Chapter 2). Anaerobic metabolism leads to **local acidosis** (lactate accumulation), which leads to **muscle soreness.** The low pH may cause release of free Fe ions from transferrin.

Effect of exercise on the immune system.

An important source of ROMs are the phagocytes (Chapter 3). During exercise **tissue damage and local inflammation** are observed. The leukocyte activation occurs from damage to muscle fibers and connective tissues during muscle contraction. Downhill running (**eccentric exercise**) is believed to exert greater tension to muscle fibers and adjacent connective tissue than uphill running (**concentric exercise**) and should therefore lead to more damage and inflammation (see Eberhardt, 2000).

Long-term intensive training does not affect the quantitative composition of white blood cells. However the **neutrophil bactericidal activity** was lower in sportsmen than in the untrained controls at rest. The neutrophils of trained individuals produced less ROMs in response to a stimulus. These results indicate that trained individuals are more susceptible to common infections (viral, bacterial and parasitic) (Jokl, 1974).

The reduced neutrophil activity in trained athletes may be due to reduced tissue damage upon exercise, since training strengthens muscle fibers and diminishes muscle damage.

During exercise the total number of leukocytes increases proportional to the duration of the exercise. The exercise however not only increased the total number of leukocytes, but also increased their activity. Thus, the response of polymorphonuclear leukocytes to physical exercise is similar to their response to infection. The damaged tissue is removed by the leukocytes (Chapter 3).

Epidemiological studies have found a correlation between elevated leukocyte counts and the incidence and mortality for myocardial infarction. **Neutrophils** play an important role in the pathogenesis of vascular injury. Advanced **atherosclerosis** has indeed been found at autopsies of men who died from coronary artery disease during strenuous exercise.

Antioxidant defenses during exercise

The increased formation of ROMs during exercise is counteracted by antioxidant defenses (Chapter 4). The induction of antioxidant defenses in skeletal muscle is linked to the increase in radical production $(O_2\cdot^-)$ in the mitochondria associated with exercise.

As I have pointed out before the antioxidant defenses are not evenly distributed throughout the body. They are produced where they are needed most. Liver and heart have higher antioxidant levels than skeletal muscle. Liver has the highest antioxidant level in the human body. During exercise the oxygen consumption in heart and liver increases only modestly (4-5 times), whereas O_2-consumption in muscles increases up to 20-fold. Because radical production is proportional to oxygen uptake, the liver and heart may be subjected to only moderate oxidative stress during exercise.

Another defense against ROM-induced tissue damage during exercise is the elimination of **Cu and Fe ions** in the sweat of athletes (Gutteridge et al. 1985). It has been suggested that excretion reduces the risk of lipid peroxidation. However I have already pointed out several studies, which showed increased levels of lipid peroxidation (ethane and pentane formation) during exhaustive exercise. These results clearly indicate that the defensive mechanisms (induction of antioxidant enzymes and excretion of Cu and Fe ions in the sweat) are not sufficient to counteract increased damage by ROMs. Vitamin E reduced, but did not abolish exercise-induced lipid peroxidation. Remember: we always have competition between ROMs reacting with biomolecules and with antioxidants.

How do our cells cope with vigorous exercise over a lifetime? In order to meet the increased need for energy, our cells have to build new power plants. The muscle cells of trained athletes have indeed been found to contain more mitochondria (Davies et al., 1981).

The pros and cons of exercise

There are both beneficial and deleterious effects of exercise training. It is, like everything else in medicine, a question of **balance**. I have discussed many examples, such as the case of SOD, ascorbic acid, uric acid, Fe ions, cholesterol and oxygen. We are dealing with a complex dynamic system, where the initial conditions are not precisely defined. Considering biochemical individuality and the huge number of people exercising, it is not too surprising to find many individuals, which show deleterious effects upon exercise.

We also have to consider **environmental factors**. Ozone considerably increases pulmonary damage during exercise. Although the absolute risk of cardiovascular complications of exercise is low, the risk of sudden death and cardiac arrest seems to be higher during exercise than at rest. It has been estimated that the risk of sudden death is 4.5 times higher in prolonged cross-country skiing than during the rest of the day. Exercise consisting of at least 10 miles brisk walking or jogging per week, combined with **diet (low fat, low cholesterol)** has been shown to have a beneficial effect on the **HDL/LDL ratio** (Stefanik et al., 1998). The effect was more pronounced in men than in women. Obviously men and women are not created equal.

A recent study, which compared walking with vigorous exercise, was carried out on 72,488 female nurses (Manson et al. 1999). The data indicate that brisk walking (about three or more hours per week) and vigorous exercise are associated with substantial and similar reductions in the incidence of coronary events. This result, of course, has been known to Puerto Ricans for centuries (*no hay nada que dure mas, que un vago bien cuidado*).

Everything in excess is bad. Remember Paracelsus: everything is poison, it only depends on the dose. Voluntary exercise did improve survival in rats (i.e. more attained old age), but did not result in an extension of lifespan.

Another important factor in considering the pros and cons of exercise is the age of the individual. Experiments with rats have shown that there exists a **"Threshold Age"**, above which it is disadvantageous for the rat to commence an exercise program (Edington et al., 1972). The older rats do not adapt to the increased oxidative stress. The moral of the story: Old men or women, who never exercised before, should not run a marathon!

Alcohol consumption

The consumption of alcoholic beverages represents an excellent example of the wisdom expressed by Paracelsus. High alcohol consumption leads to numerous diseases, like **cirrhosis of the liver**, and a variety of **cancers** (oral, pharyngeal, laryngeal, esophageal and liver). However alcohol taken in moderate amounts (1-3 drinks per day) has been claimed to lower the risk of **coronary artery disease (CAD)**.

Epidemiological studies have shown that consumption of alcoholic beverages is related to cancer in humans. However it is less clear if the compound responsible is ethanol. Alcoholic beverages contain aldehydes, N-nitrosamines, phenols, flavonoids, tannin, acrolein and pesticide residues. N-nitrosamine present in some beers are known to be carcinogenic. **Flavonoids**, on the other hand, are effective antioxidants. **Ethanol** is actually an efficient **antioxidant** in its own right. This antioxidant (radical scavenging) property is commonly used for the identification of hydroxyl radicals via **spin trapping** (Chapter 2, and Eberhardt, 2000).

The mitochondrial respiratory chain produces $O_2^{\cdot-}$ and H_2O_2 and the formation of $O_2^{\cdot-}$ in mitochondria is considerably elevated after acute alcohol ingestion. However alcohol consumed in moderate amounts also has some beneficial effects. Numerous studies have shown that moderate alcohol consumption protects against **coronary artery disease (CAD)**. Another beneficial effect of moderate alcohol consumption is the increase in **HDL-cholesterol** (Srivastava et al., 1994), which may be of particular significance with regards to **atherosclerosis**. The only other method for increasing HDL-cholesterol is through mild exercise or walking combined with diet.

Another mechanism against atherogenesis may involve **flavonoids**, which are excellent antioxidants. Flavonoids are present in plants, fruits, vegetables and beverages, such as tea, coffee, beer and especially red wine.

There is an interesting relationship between alcohol consumption and melatonin levels (Reiter and Robinson 1995).

It has been reported that alcohol consumed in the early evening lowers melatonin levels, but alcohol consumed at bedtime increases melatonin at least for the next few hours. A glass of red wine at bedtime may therefore have the dual beneficial effect of increasing both the level of flavonoids and melatonin.

Diet

> *Thy food shall be thy medicine*
> *Hyppocrates*

The recommendations about diet are constantly changing. Sometimes confusion reigns. This is not surprising for several reasons. Science advances, new ideas come up, but we also have to remember that science is a typical human endeavor. Humans are complex systems with complex motivations. They may be compared to a programmed computer and therefore they have preconceived ideas. The program also contains memes, put there by our environment. This is why some scientists are not only motivated by curiosity and the search for truth, but also by power, fame, and money. First we were told not to eat butter, then it turned out that margarine was even worse. Soy oil is supposed to be good, if you don't worry about a report claiming that it atrophies the brain. What about chocolate? Chocolate was thought to be bad for many years, until some scientists discovered that it contained antioxidants. This may be so, but chocolate also contains lots of fat and carbohydrates, which are all bad for us. So, instead of eating chocolate we would be better off eating fruits and vegetables. On the other hand, by eating chocolate we not only support a huge industry, but also child slave labor in Africa. I hope to make sense out of this confusion by pointing out some basic chemistry, which I have discussed in previous chapters.

The accumulated damage to lipids, proteins and DNA by ROMs during a lifetime is responsible for aging, cancer and other degenerative diseases of old age. It appears possible that we can slow the aging process and increase life span by decreasing oxidative stress. This can be accomplished by lowering ROM formation (**metabolic rate**) or by increasing the levels of endogenous antioxidants (SOD, CAT, GSH-Px, uric acid, carnosine, melatonin) or by adding dietary antioxidants (vitamins A, C, E and selenium).

Some studies on rats have suggested that **calorie restriction (CR)** as well as **protein restriction (PR)** markedly extends life span and inhibits carcinogenesis. CR and PR decrease lipid peroxidation, lipofuscin accumulation, protein oxidation and DNA damage. DNA damage was more extensive in individuals eating few fruits or vegetables. Specific metabolic rate in many species has been inversely correlated with life span and oxidative damage to DNA. It has been suggested that animals with **high metabolic rates** produce more ROMs, causing greater damage to macromolecules (Adelman et al. 1988). This is why mice have a shorter lifespan than humans and why there are so many rodent carcinogens (Ames and Gold, 1990). CR and PR not only decreases ROM formation, but also increases the levels of endogenous antioxidants, such as SOD, CAT, and GSH-Px (see Eberhardt, 2000).

In an oxygen-respiring organism, cellular homeostasis is con-tinuously challenged by ROMs In an old organism this threat could be more serious because of the increased ROM production and deteriorating defenses during aging. It was suggested that food restriction slows the aging process.

CR reduces membrane damage. It was observed that unlike aged rats on a regular diet, rats on a hypocaloric diet presented brain membrane characteristics similar to those of young rats.

This is interesting in relation to the deterioration of memory in aged rats. A life-long **hypocaloric diet** acts to prevent age-related memory deficits (Pitsikas and Algeri, 1992).

A **low protein diet** was found to have a beneficial effect on tumor development. It was shown that development of aflatoxin B_1-induced hepatic foci were markedly decreased by low protein feeding. Foci development, tumor incidence, tumor size, and the number of tumors per animal were markedly reduced while the time to tumor emergence was increased with low protein feeding (Youngman and Campbell, 1992). A study on diet, lifestyle and mortality in 65 counties in China indicated that age at menarche is significantly prolonged by a low protein diet. In view of all the available evidence for the damaging effects of a high protein diet it is indeed amazing that some medical gurus (Atkins diet) still recommend a high protein diet

Another important part of a good diet is the **low intake of fats**. Epidemiological studies by Willett and coworkers (Willett et al. 1993 and Hu et al. 1997) have indicated that the intake of *trans* **fatty acids** increases the risk for coronary artery disease among women. As I have already discussed (Chapter 1, p.15) most naturally occurring fatty acids are in the *cis* configuration. *Trans* fatty acids have been introduced in to our food supply by partial hydrogenation of unsaturated vegetable oils to produce margarine and vegetable shortening. *Trans* fatty acids are estimated to constitute 5-6% of dietary fat consumed in the USA, but the amounts vary widely, depending on personal habits. Willett et al. (1993) have shown that *trans* fatty acids **increase LDL** (low density lipoprotein) and **decreases HDL** (high density lipoprotein), while only slightly raising total cholesterol. It has been suggested (Hu et al. 1997) that replacing saturated and *trans* unsaturated fats with unhydro-genated monounsaturated and polyunsaturated fats is more effective in preventing coronary artery disease in women than reducing overall fat intake.

A diet supplemented with argine (the precursor of NO) has been shown to lower the risk of atherosclerosis in hyper-cholesterolemic rabbits (Cooke et. al., 1992). This is due to the protective effect of NO in the oxidative modification of LDL-cholesterol (see under atherosclerosis).

Among all the hoopla about the dangers of saturated fats it has often been forgotten, that not only saturated fats, but unsaturated fats as well are bad for you. Unsaturated fatty acids are oxidized to hydroperoxides, which interact with metal ions (Chapter 2) or with Vitamin C to give peroxyl radicals, thus damaging DNA (Lee et al. 2001). I have already discussed these reactions under pros and cons of ascorbic acid (pp. 100-104) So, the recommendation to replace saturated fats with unsaturated fats is not necessarily good advice. Less total fat is obviously better!

ROM-induced oxidative damage is not only affected by caloric restriction and protein restriction, but also by the **fat content** of the diet. DNA damage in nucleated peripheral blood cells in a group of women with a high risk for **breast cancer**, was found to be three-fold higher in the group on a non-intervention diet, compared to the low fat diet group (Djuric et al. 1991). A high fat diet has been associated with increased breast cancer risk (Schatzkin et al. 1989, Howe et al. 1990) and the increased DNA damage may provide the mechanistic link.

From all the above evidence, we may appreciate the wisdom expressed by Hyppocrates.

Psychological stress

In recent years books on the topic **mind over matter** appear to be proliferating. Most of these books are not based on solid science. This may be due to the complexity of science. The general public has great difficulty understanding scientific concepts and can impossibly keep up with the explosive growth of science. It is a lot easier to seek refuge in meditation or religion. These books dealing with the topic mind over matter, therefore have wide appeal, especially when these ideas are expressed by famous people, some even with scientific credentials. However, we should remember, that even famous people have the right to make fools of themselves.

The association between personality and the risk of developing **coronary artery disease** has been extensively investigated. It has been shown that certain behaviour patterns (Type A) was related to an increased risk to develop coronary heart disease. I have already discussed (under atherosclerosis) that mental stress causes a paradoxical **vasoconstriction** leading to ischemia. A hostile personality is indeed a risk factor (among 245 more) for coronary artery disease.

A causal relationship between stress and diseases, which have a long-term development such as cancer, is difficult to prove. Cancer is caused by numerous factors and its development is modulated by genetic factors, by age and the state of the immune system. It is therefore not too surprising to find lots of controversy concerning **the stress-cancer link**.

What is the scientific evidence linking psychological stress to disease. We have to show that stress causes the increase of a biological marker for oxidative stress (Chapter 2). Adachi et al. (1993) have demonstrated **increased 8-OHdG formation** in the livers of emotionally stressed rats. These results of Adachi et al. may provide a connection between stress and cancer (8-OHdG is mutagenic). Although the results show repair after cessation of the stress, the repair system may not be able to cope with continued psychological stress. Psychological stress has indeed been shown to lead to decreased DNA repair in human blood cells (Kiecolt-Glaser et al., 1985). Human subjects, which could not sleep all night, showed a conside-rable increase in lipid peroxidation products (aldehydes) in the urine (Kosugi et al., 1994).

Another difficulty in establishing a causal link between stress and disease is the fact that each human is an individual with different initial conditions. What constitutes stress or excess stress varies for each individual. Obviously not all men are created equal!

The negative effects of stress on the efficiency of the im-mune system are well documented (Solomon, 1987). Stress is known to reduce the activity of natural killer cells. Cancer may not be caused by stress, but may develop as a consequence of a deficient immune system.

Summary

Some of the guidelines for a longer and healthier life are: a low fat (low in *trans* fats), low protein, low calorie diet with lots of fruits and vegetables. Moderate exercise, moderate alcohol consumption (especially recommended two glasses of red wine per day), no smoking and avoidance of psychological stress.

Chapter 6

EPIDEMIOLOGY

> *In a universe of selfish genes, blind physical forces and genetic replication, some people are going to get hurt, others are going to get lucky, and you won't find any rhyme or reason for it.*
>
> *Richard Dawkins in "The blind watchmaker".*

Can we increase our life expectancy by **antioxidant supplementation** or can we decrease the likelihood for the occurrence of various diseases? Statistics only works with large numbers. These studies in large human populations take many years to complete at great expense, and in the end we find out what we (or at least some of us) already knew. Taking vitamin supplements does not protect you (as an individual) from dying of cancer or heart disease. As I pointed out in Chapter 4, antioxidants can never offer complete protection against ROMs. We always have a **competition** between ROMs reacting with biomolecules and antioxidants. Large scale studies on antioxidant supplementation have, therefore, shown less than dramatic effects. Lowering the incidence of a disease by 20-30% is hailed as a medical breakthrough. Obviously 70-80% of people still get sick and die. Maybe you are one of the lucky ones. Life obviously is a gamble, some win, some loose.

Why should this be so? Humans are complex dynamic systems, where the initial conditions are different for each individual. Drugs are accompanied with pamphlets listing a huge number of side effects, which may or may not occur. The complexity and different initial conditions are the principal reasons why medicine is an art and not yet a science.

We are a pill popping society. Many people believe that a physician who doesn't give them an expensive prescription is no good. Remember: everything is poison, it just depends on the dose. This dose is different for each individual. Any abnormal symptoms we experience (headache, upset stomach, fever, sleeplessness etc.) we treat with a pill. We treat the symptoms, rather than the cause. Instead of just resting and adjusting our lifestyle, we are taking antipain, antiacids, antidepressants, antibiotics and other anti-whatever. This is, of course, greatly encouraged by the pharmaceutical industry, which puts these **viruses in our minds** (Dawkins, 2003). Every night at dinner time the News is interrupted with Commercial after Commercial promoting all kind of drugs. Presently antiacids are very popular. They are advertised with the warning, that some patients may experience nausea, vomiting, diarrhea or intestinal cramps. Just keep on eating any food you want and keep on taking our antiacid pill. Taking any kind of medication to treat disease should be the last recourse, as was already recognized by Hyppocrates: *Thy food shall be thy medicine*.

This statement by Hyppocrates makes good sense in light of recent ideas discussed by Crawford and Marsh (1995). According to their theory evolution has been guided by the presence or absence of certain compounds in our food supply. I have already discussed some examples in Chapter 4 (Vitamin A and E). It is therefore not surprising, that our food will have a profound effect on our well being. The fact that we humans eat a great variety of foods, makes the species *homo sapiens* the most successful species among the primates.

Another example of our pillomania is the overuse of **antibiotics**, which in most cases (viral infections) does no good. It would be better to rest and eat chicken soup. The overuse of antibiotics shows us **evolution in action.** Many bacteria have developed resistance against antibiotics, and someday we may be left completely defenseless against bacterial infections.

Since **cancer** and **coronary artery disease** are the two most important causes of death in the USA and other Western countries, the effect of vitamin supplementation on the frequency of these diseases has been extensively investigated . Some investigators are convinced that their studies show a beneficial effect of **Vitamin E** on CAD, whereas others are less than convinced. **Vitamin C** supplementation showed no effect on CAD (reviewed by Byers, 1993). In one study the beneficial effects were limited to a sub-group receiving very high doses of Vitamin E. Since nothing is known about the possible long-term effects of Vitamin E, some scientists warn: hold the Vitamin E (Steinberg 1993). I have already discussed the **prooxidant effects** of Vitamin E and C in presence of metal ions. Because of the prooxidant effect of Vitamin C some researchers have discussed the possible role of Vitamin C as a slow-acting carcinogen (Halliwell 1994). Excess Vitamin C is excreted, but it is also catabolized to **oxalic acid**, which combines with **calcium ions (Ca^{2+})** to form the insoluble calcium oxalate, leading to kidney and bladder stone formation. Stones in kidney or bladder may irritate the epithelial cells and initiate carcinogenesis (Chapter 5).

A large scale nutritional study (3968 men and 6000 women, aged 25-74 over a period of 19 years) examined the intake of **vitamins A, C and E** and the risk for **lung cancer** (Yong et al. 1997). There was no additional protective effect of vitamin supplementation, beyond that produced by diet alone. Another study examined the effectiveness of Vitamin E and β-carotene in preventing lung cancer in male smokers (Heinonen et al. 1994). Vitamin E showed no reduction in lung cancer rates and β-carotene showed a small increase (18%). On the other hand fewer cases of prostate cancer were diagnosed in patients receiving Vitamin E. The moral of the story: you cannot counteract a bad habit, like smoking, with antioxidant supplementation!

As I have discussed (Chapter 4), the most important antioxidants are produced by our own metabolism. Some of these antioxidants are: SOD, CAT, and GSH-Px. In order for our bodies to synthesize these enzymes, we need certain compounds. These compounds include **selenium (Se)**, for the synthesis of Glu-Px, **Zn, Cu, Fe and Mn** for the synthesis of SOD and CAT. These metal ions are essential for the catalytic activity of these enzymes. Ecological studies in the USA have shown an inverse correlation between cancer mortality rates and Se in locally grown **food crops** (Clark et al., 1991) as well as with **dietary Se** intake (Hocman, 1988). There is increasing epidemiologic evidence indicating that Se plays a prominent role in cancer prevention. Some interesting epidemiological studies have been carried out in China. People in certain regions of China have the lowest Se intake in the world and at the same time have very high rates of lung, esophagus, stomach and liver cancers (Yu et al., 1985). Since GSH-Px removes H_2O_2, any deficiency in Se will cause a decrease in the GSH-Px concentration and thus an increase in the steady state concentrations of H_2O_2. This changes the prooxidant antioxidant balance and leads to pathological changes.

In a recent study of cancer prevention by **Se supplementation**, it was shown that Se did not lower the risk of skin cancer (basal cell and squamous cell carcinomas), but showed profound effects on the incidence of **lung cancer, colorectal cancer** and **prostate cancer** (Clark et al. 1996). This different response of different organs is not surprising since it is well known that the antioxidant defenses are not evenly distributed throughout the body. Antioxidant concentrations are higher in organs exposed to high oxidative stress, especially the lung. Selenium and the enzyme glutathione peroxidase derived from Se are therefore expected to have a more profound effect on lung tissue than on skin.

Supplementation with Se, Cu, Fe, Zn, Mn, does, of course, no good if the genes, specifying the synthesis of SOD, CAT or GSH-Px are defective. In any disease, caused by a defective gene (as in neurodegenerative disorders), antioxidant supplementation will do no good. It will not cure the cause, but (at best) may alleviate the symptoms.

As I discussed under melatonin (Chapter 4) the biosynthesis of the antioxidant melatonin requires tryptophan and calcium ions (Ca^{2+}). Therefore **high Ca^{2+} intake** should prevent carcinogenesis, contrary to epidemiological results on the relationship between high milk and dairy product consumption and prostate cancer (Giovanucci et al. 1998). We have another **paradox**. It is a question of balance.

Inhibition of chemical induced carcinogenesis by **melatonin** has been demonstrated in rats (Chapter 4) However, the results on a small number of humans, to study prevention of cancers (melanomas, prostate, lung and liver) or the shrinking of existing tumors, has been less than overwhelming (Reiter and Robinson 1995). You may be better off praying!

Our health and longevity is determined by **genes, environment** and **life style**. It is of great interest to know the relative contribution of genes (nature), environment and lifestyle (nurture) to our health and longevity. In epidemiological studies these factors are not easy to separate. A recent study on a huge number of twins in Scandinavia tries to answer this question for cancer and heart disease (Lichtenstein et al, 2000). The authors of this study claim, that the cause of many cancers depend more on the environment and lifestyle than on he redity. For a response see Hoover (2000).

All cancers are ultimately genetic, since they are caused by a damaged gene. This damaged gene may be inherited or may be produced by environmental toxins, diet or lifestyle (smoking, drinking) The ratio genetic/environmental most likely is different for each disease. The authors of the twin study have indeed observed different nature/nurture ratios for different types of cancer. The estimates by Lichtenstein et al. showed that 73 % of the causation of **breast cancer** is environmental and 27% heritable. This result is consistent with studies of migrant groups. Breast cancer among women who have recently immigrated to the USA from rural Asia are similar to those in their homeland and about 80% lower than the rates among third-generation Asian-American women, who have rates similar to white women in the US.(Ziegler et al., 1993) **Prostate cancer** showed the greatest genetic component (42%). which is at variance with other data. Prostate cancer shows marked international variation, and the risk of migrant groups tends to rise towards the level in the adopted country, indicating a substantial environmental component of the risk of this cancer. One of these components is the high intake of Ca^{2+} through high milk and dairy product consumption (**the melatonin-calcium-prostate paradox**).

Another controversial subject is the value of **estrogen** or the estrogen-progestin combination for the well being of post-menopausal women. Despite an extensive literature on the subject, the jury is still out. For a discussion see Nabel, 2000, Herrington et al. 2000, Hu et al. 2000. There is some evidence that estrogen prevents hot flashes, vaginal dryness and possibly osteoporosis and coronary artery disease. However estrogen is also a known **carcinogen** (Henderson et al. 1988), increasing the risk for **breast cancer**.

A recent study, which included over 5000 women presented strong evidence that the estrogen-progestin combination therapy increased the breast cancer risk by 24% (Ross et al. 2000, and Pike and Ross, 2000). This therapy was called **"combination hormone replacement therapy"** or CHRT. The authors claim that the adverse effects of hormone replacement therapy on the breast may outweigh the beneficial effect on the endometrium. However, the data and its interpretation by Ross et al (2000) was strongly criticized by Archer et al. (2000). In response Pike and Ross (2000) state, and I quote: The rather intransigent attitude of Dr. Archer and colleagues regarding their patients is especially unfortunate. To choose not to inform patients of the mounting, and now considerable, evidence that CHRT causes a large increase in breast cancer risk, is a real disservice.

I tend to agree. The risk for coronary artery disease can be lowered by other means, such as changes in **diet and lifestyles**. A recent study examining the connection between diet and lifestyles in women and CAD was published in the New England Journal of Medicine (Hu et al. 2000). These authors followed 85 941 healthy women (34 to 59 years of age) from 1980 to 1994 in the Nurses' Health Study. Diet and lifestyle factors were assessed through questionaires. These authors concluded that a reduction in **smoking**, and an improved **diet** (low fat, especially low in *trans* fatty acids, low cholesterol, high fiber and high in fruits and vegetables) can considerably decrease the incidence of CAD (*thy food shall be thy medicine*). On the other hand, an increasing prevalence of **obesity**. appears to have slowed the decline in the incidence of CAD. Hu et al. (2000) estimated that the reduction in smoking (by 41%) explained a 13 percent decline in the incidence of coronary artery disease, improvement in diet explained a 16 percent decline, post-menopausal hormone use (increased 175%) explained a 9 percent decline, whereas the increase in obesity (increase of 38%) explained an 8 percent increase in the incidence of CAD over the study period from 1980 to 1994. In other words keeping your weight down, through proper diet and moderate exercise, such as brisk walking for three hours or more per week (Manson et al. 1999, see p. 150), will give women the same protection against CAD as **hormone replacement therapy**, but without the possible risk for **breast cancer**. Women with a family history for breast cancer should be especially careful in taking estrogen supplementation. A government advisory panel has recently (2000) recommended that estrogen be added to the list of cancer-causing chemicals.

Testosterone decreases life expectancy, and it increases red blood cells, causes blood clots and heart attacks. Recently a testosterone patch has become available to increase libido and physical strength. Men using this patch are obviously making a pact with the devil.

EPILOGUE

I hope that in this book I have succeeded in giving the reader a glimpse of the complexity of life in an oxygen atmosphere. Aging and many diseases of old age are an inevitable consequence of life in an oxygen atmosphere. Obviously with every breath we take, we get closer to the grave. Our fate is intimately linked to the chemistry of reactive oxygen metabolites. During billions of years of evolution we have adapted to the negative effects of oxygen metabolism. To assume that we can outsmart evolution by antioxidant supplementation is foolhardy. Antioxidants may only serve as a bandaid against ROM induced damage. However, the studies on **twins** and the epidemiological data on **diet and lifestyles** indicate, that we are not completely at the mercy of our genes. We can to some extent decrease the likelihood of contracting certain diseases by proper lifestyles and thus may increase our lifespan to a limited extent. Statistics shows a steady increase in the **mean life span** (life expectancy) in the USA during the last century. At present the life expectancy in the USA does not exceed 75 years, far below the estimated maximum of 120 years (Schulz-Aellen, 1997). During the last century medical science has made tremendous progress. One may, therefore, be tempted to ascribe our increased life expectancy to better medical care. However, most of it is due to better nutrition (thy food shall be thy medicine) and public health measures (safe water supply and hygiene) and antibiotics. The life expectancy (mean life span) was much lower in previous centuries, because many people died prematurely of infectious diseases (Mozart, Chopin, Schuman, Gauguin).

Medical science has advanced most dramatically in the area of diagnosis and mechanical repair, and much less so in finding cures. Diagnosis and surgery, of course, require the right tool for the job. As science progresses these tools get more and more expensive. We are eventually reaching the point of diminishing returns and have to deal with ethical questions. In the future, we will be able to repair defective genes and thus eliminate certain diseases and even increase our life expectancy. However, until that time comes, don't hold your breath.

LITERATURE

As mentioned in the introduction, this book is a simplified and condensed version of a previous book: Manfred K. Eberhardt, Reactive Oxygen Metabolites, Chemistry and Medical Consequences, CRC Press, Boca Raton, FL, 2000. The interested reader can find additional references in this text.

Books

Ball, Ph., *Life's Matrix,* Farrar, Straus and Giroux, New York, 1999.

Barrow, J. D., *The World within the World,* Oxford University Press, Oxford and New York, 1988, p. 4.

Biadasz Clerch, L. and Massaro, D. J., in Preface to *Oxygen, Gene Expression, and Cell Function.* L. Biadasz Clerch and D. J. Massaro, eds., Marcel Dekker, New York, 1997.

Bova, B., *Immortality. How science is extending your life span and changing the world,* Avon Books, New York, 1998.

Brown, G., *The Energy of Life. The science of what makes our minds and bodies work,* The Free Press, New York, 2000.

Capra, F., *The Tao of Physics,* Third Edition, Shambhala, Boston, 1991, p.114.

Casti, J. L., *Paradigms regained. A further exploration of mysteries of modern science,* Harper Collins Publishers Inc., New York, 2000.

Crawford, M. and Marsh, D., *Nutrition and Evolution,* Keats Publishing, Inc., New Canaan, Connecticut, 1995.

172

Czapski, G. in *Oxygen and Oxy-Radicals in Chemistry and Biology,* M. A. J. Rodgers and E. L. Powers, eds., Academic Press, New York, 1981, pp. 236-239.

Davies, P., *The fifth miracle. The search for the origin and meaning of life,* Simon and Schuster, New York, 1999

Dawkins, R., *a devil's chaplain,* Houghton Mifflin Company, Boston, New York 2003 and references cited therein.

Dyson, F., *Origins of Life. Second Edition,* Cambridge University Press, Cambridge, UK, 1999.

Eberhardt, M. K., *Reactive Oxygen Metabolites, Chemistry and Medical Consequences,* CRC Press, BocaRaton, FL , 2000

Emsley, J., *Molecules at an Exhibition. Portraits of intriguing molecules in everyday life,* Oxford University Press, New York, 1998.

Fox, S., *The Emergence of Life. Darwinian evolution from the inside.* Basic Books, Inc. Publishers, New York, 1988.

Fridovich I., contra J. A. Fee, A discussion can be found in: *Oxygen and Oxy-Radicals in Chemistry and Biology,* M.A. J. Rodjers and E. L. Powers, eds., Academic Press, New York, 1981, pp. 197-239.

Hesse, H., *The Glass Bead Game (Magister Ludi),* Henri Holt, and Company, New York, 1966. Translated by Richard and Clara Winston (p. 16 and 379).

Hoffmann, R., *The Same and not the Same,* Columbia University Press, New York, 1995.

Hofstadter, D., *Gödel, Escher and Bach,* Basic Books, Inc., New York, 1979.

Lederman, L. with Teresi, D., *The God particle,* Bantam Doubleday Dell Publishing Group, Inc., New York, 1993.

Lovelock, J. E., *Gaia: A New Look at Life on Earth,* Oxford University Press, Oxford, 1979.

Margulis, L., *Symbiosis in Cell Evolution,* Freeman and Co., San Francisco, 1981.

Margulis, L. and Sagan, D., *What is Life?,* Simon and Schuster, New York, 1995.

Margulis, L. and Sagan, D., *Slanted Truth. Essays on Gaia, Symbiosis and Evolution,* Copernicus, Springer, New York, 1997.

v. Neumann, J., The general and logical theory of automata. Lecture given 1948. In *Cerebral mechanism in behaviour-* The Hixon Symposium, ed. L.A. Jeffres, pp. 1-41, John Wiley, New York, 1951.

Pascale, R. T. Millemann, M. and Gioja, L., in *"Surfing the edge of chaos",* Crown Business, 2000

Pauling, L., *The Nature of the Chemical Bond,* Cornell University Press, Ithaca, N.Y., 1940.

Pauling, L., *Vitamin C and the Common Cold,* W. H. Freeman, San Francisco, 1970.

Pollack, R., *Signs of Life. The language and meanings of DNA,* Houghton Mifflin Company, Boston, New York, 1994.

Reiter, R. and Robinson, Jo, *Melatonin,* Bantam Books, New York, 1995

Rhodes, R., *Deadly Feasts. Tracking the Secrets of a terrifying new Plague,* Simon and Schuster, New York, 1997.

Schulz-Aellen, M.-F., *Aging and human longevity,* Birkhäuser, Boston, 1997.

von Sonntag, C., *The Chemical Basis of Radiation Biology,* Taylor and Francis, New York, 1987.

Szent-Györgyi, A., *The Living State and Cancer,* Marcel Dekker, New York, 1978.

Varmus, H. and Weinberg, R. A., *Genes and the Biology of Cancer,* Scientific American Library, a Division of HPHLP, New York, 1993.

Waldrop, M. M., *Complexity. The emerging science at the edge of order and chaos,* Simon and Schuster, New York, 1992.

Weinberg, S., *Dreams of a Final Theory, The scientific search for the ultimate laws of Nature,* Vintage Books, a division of Random House, Inc., New York, 1992.

Articles

Adachi, S., Kawamura, K. and Takemoto, K., Oxidative damage of nuclear DNA in liver of rats exposed to psychological stress, *Cancer Res.,* 53, 4153, 1993.

Adams, C. W. M., Perivascular iron deposition and other vascular damage in multiple sclerosis, *J. Neurol. Neurosurg. Psychiatr.* 51, 260, 1988.

Adelman, R., Saul, R. L. and Ames, B. N., Oxidative damage to DNA: Relation to species metabolic rate and life span, *Proc. Natl. Acad. Sci. USA,* 85, 2706, 1988.

Alessio, H. M., Exercise-induced oxidative stress, *Med.Sci. Sports Exer.,* 25, 218, 1993.

Allen, R. C., Stjernholm, R. L. and Steele, R.H., Evidence for the generation of an electronic excitation state in human polymorphonuclear leukocytes and its participation in bactericidal activity, *Biochem. Biophys. Res. Commun.,* 47, 679, 1972.

Ames, B. N., Cathcart, R., Schwiers, E., Hochstein, P., Uric acid provides an antioxidant defense in humans against oxidant and radical-caused aging and cancer; a hypothesis, *Proc. Natl. Acad. Sci. USA* 78, 6858, 1981.

Ames, B. N., Dietary carcinogens and anti-carcinogens (oxygen radicals and degenerative diseases), *Science,* 221, 1256, 1983.

Ames, B. N. and Gold, L. S., Chemical Carcinogenesis: Too many rodent carcinogens, *Proc. Natl. Acad. Sci. USA,* 87, 7772, 1990.

Ames, B. N. and Shigenaga, M. K., Oxidants are a major contributor to cancer and aging, in *DNA and Free Radicals,* B. Halliwell, O. I. Aruoma, eds, Ellis Horwood, Chichester, England, 1993, pp. 1-15.

Arnold, W. P., Mittal, C. K., Katsuki, S. and Murad, F., Nitric Oxide activates guanylate cyclase and increases guano-sine-3',5'-cyclic monophosphate levels in various tissue preparations, *Proc. Natl. Acad. Sci. USA,* 74, 3203,1977.

Archer, D. F., Bush, T. and Nachtigall, L. E., Re: effect of hormone replacement therapy on breast cancer risk: estrogen versus estrogen plus progestin, *J. Natl. Cancer Inst.,* 92, 1950, 2000.

Babior, B. M. and Crowley, C. A., Chronic granulomatous disease and other disorders of oxidative killing by phagocytes in *The metabolic basis of inherited disease,* Fifth Edition, J. R. Stanbury, J. B. Wyngaarden, D.S. Fredrickson, J. L. Goldstein, M. S. Brown, eds., McGraw Hill, New York, 1983 pp.1956-1984.

Babior, B. M., Kipnes, R.S. and Curnutte, J. T., Biological defense mechanisms: The production by leukocytes of superoxyde, a potential bacteriocidal agent, *J. Clin. Invest.,* 52, 741, 1973.

Bagasra, O., Michaels, F. H., Zheng, Y. M., Bobroski, L. E., Spitsin, S. V., Fu, Z. F., Tawandros, R. and Koprowski, H., Activation of the inducible form of nitric oxide synthase in the brains of patients with multiple sclerosis, *Proc. Natl. Acad. Sci. USA,* 92, 12041, 1995.

Barnes, P. J., Reactive oxygen species and airway inflammation, *Free Radical Biol. Med.,* 9, 235, 1990.

176

Behl, C., Davis, J., Cole, G. M. and Schubert, D., Vitamin E. Protects nerve cells from amyloid β protein toxicity, *Biochem. Biophys. Res. Commun.*, 186, 944, 1992.

Behl, C., Davis, J., Lesley, R. and Schubert, D., Hydrogen peroxide mediates amyloid β protein toxicity, *Cell,* 77, 817, 1994.

Bielski, B. H., J. and Chan, P. C., Emzyme-catalyzed free radical reactions with nicotine-adenine nucleotides. I. Lactate dehydrogenase catalyzed chain oxidation of bound NADH by superoxide radicals. *Arch. Biochem. Biophys.*, 159, 873, 1973.

Bossaller, C., Habib, G. B., Yamamoto, H., Williams, C., Wells, S. and Henry, P. D., Impaired muscarinic endothelium-dependent relaxation and cyclic guanosine-5-mono-phosphate formation in atherosclerotic human coronary artery and rabbit aorta, *J. Clin Invest.*, 79, 170, 1987.

Bowry, V. W. and Ingold, K. U., The unexpected role of vitamin E (α-tocopherol) in the peroxidation of human low-density lipoprotein, *Acc.Chem. Res.*, 32, 27, 1999, and references cited therein.

Bö, L., Dawson, T. M., Wesselingh, S., Mörk, S., Choi, S., Kong, P. A., Hanley, D. and Trapp, B. D., Induction of nitric oxide synthase in demyelinating regions of multiple sclerosis brains, *Ann. Neurol.*, 36, 778, 1994.

Buxton, G. V., Greenstock, C. L. Helman, W. Ph. and Ross, A., Rate constants for reactions of hyhydroxyl radicals in aqueous solutions, *J. Phys. Chem. Ref. Data,* Vol. 17, No. 2, 1988.

Byers, T., Vitamin E supplements and coronary heart disease, *Nutr. Rev.* 51, 333, 1993.

Cameron, E., Pauling, L., and Leibovitz, B., Ascorbic acid and cancer: a review , *Cancer Res.*, 39, 663, 1979.

Ceballos, I., Lafon, M., Javoy-Agid, F., Hirsch, E., Nicole, A., Sinet, P.M. and Agid, T., Superoxide dismutase and Parkinson's disease, *Lancet*, 335, 1035, 1990.

Cech, T. R. and Bass, B. L., Biological catalysis by RNA, *Ann. Rev. Biochem.*, 55, 599, 1986.

Cech, T. R., RNA as an enzyme, *Scientific American,* 255, no. 5, 64, 1986.

Cech, T. R., The efficiency and versatility of catalytic RNA: Implications for an RNA world, *Gene,* 135, 33, 1993.

Cheng, K. C., Cahill, D. S., Kasai, H., Nishimura, S. and Loeb, L. A., 8-hydroxyguanine, an abundant form of oxidative DNA damage, causes G-T and A-C substitutions, *J. Biol. Chem.,* 267, 166, 1992.

Cilento, G. and Adam, W., Photochemistry and photobiology without light, *Photochem. Photobiol.,* 48, 361, 1988.

Clark, L. C., Combs, G. F. Jr., Turnbull, B. W. et al., Effects of selenium supplementation for cancer prevention in patients with carcinoma of the skin, *JAMA*, 276, 1957, 1996.

Clark, L. C., Cantor, K. P. and Allaway, W. H., Selenium in forage crops and cancer mortality in US counties, *Arch Environ Health,* 46, 37, 1991.

Cooke, J. P., Singer, A. H., Tsao, P., Zera, P., Rowan, R. A. and Billingham, M. E., Antiatherogenic effects of L-arginine in the hypercholesterolemic rabbit, *J. Clin. Invest.,* 90, 1168, 1992.

Davies, K. J. A., Packer, L. and Brooks, G. A., Biochemical adaptation of mitochondria, muscle, and whole-animal respiration to endurance training, *Arch. Biochem. Biophys.,* 209, 539, 1981.

Diehl, A. K., Gallstone size and the risk of gallbladder cancer, *JAMA,* 250, 2323, 1983.

178

Dillard, C. J., Litov, R. E., Savin, W. M., Dumelin, E. E. and Tappel, A. L., Effects of exercise, Vitamin E, and ozone on pulmonary function and lipid peroxidation, *J. Appl. Physiol. Respir. Environ. Exercise Physiol.,* 45, 927, 1978.

Dixon, W. T. and Norman, R. O. C., Free radicals formed during the oxidation and reduction of peroxides, *Nature (London),* 196, 891, 1962.

Dizdaroglu, M., Chemistry of free radical damage to DNA and Nucleoproteins, in *DNA and Free Radicals,* B. Halliwell, O. I. Arouma, edts., Ellis Horwood, New York, 1993, pp. 19-39.

Djuric, Z., Heilbrun, L. K., Reading, B. A., Boomer, A., Valeriote, F. A. and Martino, S., Effects of a low-fat diet on levels of oxidative damage to DNA in human peripheral nucleated blood cells, *J. Natl. Cancer Inst.,* 83, 766, 1991.

Dorfman, L. M., Taub, I. A. and Bühler, R. F., Pulse radiolysis studies. I. Transient spectra and reaction rate constants in irradiated aqueous solutions of benzene, *J. Chem. Phys.,* 36, 3051, 1962.

Eberhardt, M. K., Radiation-induced homolytic aromatic hydroxylation. 6. The effect of metal ions on the hydroxylation of benzonitrile, anisole and fluoro-benzene, *J. Phys. Chem.,* 81, 1051, 1977.

Eberhardt, M. K., Reaction of benzene radical cation with water. Evidence for the reversibility of OH radical addition to benzene, *J. Am. Chem. Soc.,* 103, 3876, 1981.

Eberhardt, M. K., Roman-Franco, A. A. and Quiles, M. R., Asbestos-induced decomposition of hydrogen peroxide, *Environm. Res.,* 37, 287, 1985 (received August 31, 1982!).

Eberhardt, M. K., Ramirez, G., and Ayala, E., Does the reaction of Cu$^+$ with H$_2$O$_2$ give OH radicals? A study of aromatic hydroxylation, *J. Org. Chem.*, 54, 5922, 1989.

Eberhardt, M. K., Homolytic aromatic hydroxylations via radiolysis of aqueous solutions and via metal ion - oxygen systems, *Rev. Heteroatom Chem.*, Vol. 4, pp.1-26, Shigeru Oae, editor, MYU, Tokyo, 1991.

Eberhardt, M. K. and Martinez, M. I., Radiation-induced homolytic aromatic substitution. V. Effect of metal ions on the hydroxylation of toluene, *J. Phys. Chem.*, 79, 1917, 1975.

Edington, D. W., Cosmas, A. C. and McCafferty, W. B., Exercise and longevity : evidence for a threshold age, *J. Gerontol.*, 27, 341, 1972.

Fenton, H. J. H., Oxidation of tartaric acid in presence of iron, *J. Chem. Soc.*, 65, 89, 1894.

Floyd, R. A., Watson, J. J., Wong, P. K., Altmiller, D. H. and Rickard, R. C., Hydroxyl free radical adduct of deoxyguanosine: a sensitive detection and mechanism of formation, *Free Radical Res. Commun.*, 1, 163, 1986.

Fridovich, I., Superoxide radical and superoxide dismutases, *Ann. Rev. Biochem.*, 64, 97-112, 1995.

Furchgott, R. F. and Zawadzki, J. V., The obligaory role of endothelial cells in the relaxation of arterial smooth muscle by acetylcholine, *Nature*, 288, 373, 1980.

Gomberg, M., An instance of trivalent carbon: triphenylmethyl, *J. Am. Chem. Soc.*, 22, 757, 1900.

Gajdusek, D. C., Unconventional viruses and the origin and disappearance of kuru, *Science*, 197, 943, 1977.

Gerschman, K., Gilbert, D. L., Nye, S. W., Dwyer, P. and Fenn, W. O., Oxygen poisoning and X-irradiation: a mechanism in common, *Science*, 119, 623, 1954.

Gilbert, W., The RNA world, *Nature*, 319, 618, 1986.

180

Giovannucci, E., Rimm, E. B., Wolk, A., Ascherio, A., Stampfer, M. J., Colditz, G. A. and Willett, W. C., Cal cium and fructose intake in relation to risk of prostate cancer, Cancer Res., 58, 442, 1998.

Goldstein, L. E. et al., 3-Hydroxykynurenine and 3-Hydroxy-anthranilic acid generate hydrogen peroxide and promote "-crystallin cross-linking by metal ion reduction, *Biochemistry,* 39, 7266, 2000.

Good, P. F., Werner, P., Hsu, A., Olanow, C. W. and Perl, D. P., Evidence for neuronal oxidative damage in Alzheimer's disease, *Am. J. Pathol.,* 149, 21, 1996.

Guerrier-Takada, C., Gardiner, K., Marsh, T., Pace, N. and Altman, S., The RNA moiety of ribonuclease P is the catalytic subunit of the enzyme, *Cell,* 35, 849, 1983.

Gutteridge, J. M. C., Rowley, D. A., Halliwell, B., Cooper, D. F. and Heeley, D. M., Copper and iron complexes catalytic for oxygen radical reactions in sweat from human athletes, *Clin. Chim. Acta,* 145, 267, 1985.

Haber, F. and Willstätter, R., *Ber.* 64, 2844, 1931.

Halliwell, B., Vitamin C: the key to health or a slow-acting carcinogen? *Redox Report,* 1, 5, 1994.

Harman, D., The aging process, *Proc. Natl. Acad. Sci. USA,* 78, 7124, 1981.

Harrison, J. E. and Schultz, J., Studies on the chlorinating activity of myeloperoxidase, *J. Biol. Chem.,* 251, 1371, 1976.

Hart, E. J. and Boag, J. W., Absorption spectrum of the hy-drated electron in water and aqueous solutions, *J. Am. Chem. Soc.,* 84, 4090, 1962.

Heinonen, O. P. and Albanes, D., The effect of vitamin E and beta carotene on the incidence of lung cancer and other cancers in male smokers, *N. Engl. J. Med.,* 330, 1029, 1994.

Henderson, B. E., Ross, R. and Bernstein, L., Estrogens as a source of human cancer: The Richard and Hinda Rosenthal Foundation Award Lecture, *Cancer Res.*, 48, 246, 1988.

Herrington, D. M. et al., Effects of estrogen replacement on the progression of coronary artery atherosclerosis, *New Engl. J. Med.*, 343, 522, 2000.

Hocman, G., Chemoprevention of cancer: selenium, *Int. J. Biochem.*, 20,123.1988.

Hooper, D. C., Spitsin, S., Kean, R. B., Champion, J. M., Dickson, G. M., Chaudury, I. and Koprowski, H., Uric acid, a natural scavenger of peroxynitrite, in experimental allergic encephalomyelitis and multiple sclerosis, *Proc. Natl. Acad. Sci. USA*, 95,675, 1998.

Hoover, R. N., Cancer-Nature, Nurture, or Both, *New Engl. J. Med.*, 343, 135, 2000.

Hopkins, P. N. and Williams, R. R., A survey of 246 suggested coronary risk factors, *Atherosclerosis*, 40, 1, 1981.

Howe, G. R., Hirohata, T., Hislop, T. G. et al., Dietary factors and risk of breast cancer. Combined analysis of 12 case-control studies, *J. Natl. Cancer Inst.*, 82, 561, 1990.

Hu, F. B. et al., Dietary fat intake and the risk of coronary heart disease in women, *N. Engl. J. Med.*, 337, 1491, 1997.

Hu, F. B., et al, Trends in the incidence of coronary heart disease and changes in diet and lifestyle in women, *New Engl. J. Med.*, 343, 530, 2000.

Huang, X. et al., The A β Peptide of Alzheimer's disease directly produces hydrogen peroxide through metal ion reduction, *Biochemistry*, 38, 7609, 1999.

Huie, R. E., and Padmaja, S., The reaction of NO with superoxide, *Free Radical Res. Commun.*, 18, 195, 1993.

182

Ignarro, L. J., Buga, G. M. , Wood, K. S., Byrns, R. E., and Chauduri, G., Endothelium-derived relaxing factor produced and released by artery and vein is nitric oxide, *Proc. Natl. Acad. Sci. USA,* 84, 9265, 1987.

Ihde, A. J., The history of free radicals and Moses Gomberg's contributions, *Pure and Appl. Chem.,* 15, 1, 1967.

Iyer, G. Y. N., Islam, M.F. and Quastel, J. H., Biochemical aspects of phagocytosis, *Nature,* 192, 535, 1961.

Jokl, E., The immunological status of athletes, *J. Sports Med.,* 14, 165, 1974.

Joyce, G., RNA evolution and the origin of life, *Nature,* 338 (March 16, 1989), 217.

Kanofsky, J. R., Wright, J. Miles-Richardson, G. F. and Tauber, A. I., Biochemical requirements for singlet oxygen production by purified human myeloperoxidase, *J. Clin. Invest.,* 74, 1489, 1984.

Kanter, M. M. , Lesmes, G. R. , Kaminsky, L. A., LaHamsalger, J. and Nequin, N. D., Serum creatine kinase and lactate dehydrogenase changes following an eighty kilometer race, *Eur J. Appl. Physiol.,* 57, 69, 1988.

Kasai, H., Nishimura, S., Formation of 8-hydroxy-deoxy-guanosine in DNA by oxygen radicals and its biological significance, in *Oxidative Stress, Oxidants and Antioxidants,* H. Sies, ed., pp. 99-116, Academic Press, New York 199.

Kasha, M., Introductory remarks: the renascence of research on singlet molecular oxygen, in *Singlet Oxygen,* H. H. Wasserman, R. W. Murray, eds., Academic Press, 1979.

Katsuki, S., Arnold, W., Mittal, C. and Murad, F., Stimulation of guanylate cyclase by sodium nitroprusside, nitroglycerin and nitric oxide in various tissue preparations and comparison to the effects of sodium azide and hydroxylamine, *J. Cyclic Nucleotide Res.,* 3, 23, 1977.

Kaur, H. and Halliwell, B., Evidence for nitric oxide-mediated oxidative damage in chronic inflammation. Nitro-tyrosine in serum and synovial fluid from rheumatoid patients, *FEBS Letters,* 350, 9, 1994.

Kettle, A. J. and Winterbourne, C. C., Influence of superoxide on myeloperoxidase kinetics measured with a hydro-gen peroxide electrode, *Biochem. J.,* 263, 823, 1989.

Khan, A. U., Singlet molecular oxygen from superoxide anion and sensitized fluoroescence of organic molecules, *Science,* 168, 476, 1970.

Khan, A. U. and Kasha, M., Red chemiluminescence of oxy-gen in aqueous solution, *J. Chem. Phys.,* 39, 2105, 1963.

Kiecolt-Glaser, J. K. , Stephens, R. E., Lipetz, P.D., Speicher, C. E. and Glaser, R., Distress and DNA repair in human lymphocytes, J. *Behavioral Med.,* 8, 311, 1985.

Klug, D., Rabani, J. and Fridovich, I., A direct demonstration of the catalytic action of superoxide dismutase through the use of pulse radiolysis, *J. Biol. Chem.,* 247, 4839, 1972.

Knekt, P., Reunanen, A., Takkunen, H., Aromaa, A., Heliövaara, M. and Hakuline, T., Body iron stores and the risk of cancer, *Int. J. Cancer,* 56, 379, 1994.

Knight, J. A., Reactive oxygen species and the neurodegene-rative disorders, *Ann. Clin. Lab. Sci.,* 27, 11, 1997.

Kosugi, H., Enomoto, H., Ishizuka, Y. and Kikugawa, K., Variations in the level of urinary thiobarbituric acid reactant in healthy humans under different physiolo-gical conditions, *Biol. Pharm. Bull.,* 17, 1645, 1994.

Krinsky, N. I., Singlet excited oxygen as a mediator of the antibacterial action of leukocytes, *Science,* 186, 363, 1974.

184

Kruger, K., Grabowski, P. J., Zaug, A. J., Sands, J., Gottschling, D. E. and Cech, T. R., Self-splicing RNA: autoexcision and autocyclization of the ribosomal RNA intervening sequence of tetrahymena, *Cell,* 31, 147, 1982.

Kwon, N. S., Nathan, C. F., Gilker, C., Griffith, O. Matthews, D. E., and Stuehr, D.J., L-Citrulline production from L-arginine by macrophage nitric oxide synthase, *J. Biol. Chem.*, 265, 13442, 1990.

Lampe, F. W. Field, F. H. and Franklin, J. L., Reactions of gasaeous ions. IV. Water, *J. Am. Chem. Soc.*, 79, 6132, 1957.

Land, E. J. and Swallow, A. J., One-electron reactions in biochemical systems as studied by pulse radiolysis. IV. Oxidation of dihydronicotinamide-adenine dinucleotide, *Biochim. Biophys. Acta*, 234, 34, 1971.

Lankamp, H., Nauta, W. Th., and McLean, C., A new interpretation of the monomer-dimer equilibrium of triphenylmethyl- and alkylsubstituted-diphenylmethyl radicals in solution, *Tetrahedron Lett.*, 249, 1968.

Laughton, M. J., Halliwell, B., Evans, P. J. and Hoult, J. R., Antioxidant and prooxidant actions of the plant phenolics quercetin, gossypol and myricetin, *Biochem. Pharmacol.*, 38, 2859, 1989.

Lee, S. H., Oe, T., and Blair, I. A., Vitamin C-induced decomposition of lipid hydroperoxides to endogenous genotoxins, *Science,* 292, 2083, 2001.

Lichtenstein, P., et al., Environmental and heritable factors in the causation of cancer, *New Engl. J. Med.,* 343, 78, 2000.

Long, C.A. and Bielski, B. H. J., Rate of reaction of superoxide radical with chloride-containing species, *J. Phys. Chem.,* 54, 655, 1980..

Lovelock, J. E. and Margulis, L., Atmospheric homeostasis by and for the biosphere: The Gaia hypothesis, *Tellus* 26, 2-10, 1974.

Lusis, A. J., Atherosclerosis, *Nature*, 407, 233, 2000.

Manson, JoAnn E. et al. A prospective study of walking as compared with vigorous exercise in the prevention of coronary heart disease in women, N. Engl. Med., 341, 650, 1999.

Masoro, E. J., Yu, B. P. and Bertrand, H. A., Action of food restriction in delaying the aging process, *Proc. Natl. Acad. Sci. USA,* 79, 4239, 1982.

Mathews-Roth, M. M., Pathak, M. A., Fitzpatrick, T. B., Harber, L.C. and Kass, E. H., Beta-carotene as a photoprotective agent in erythropoietic protoporphyria, *N. Engl. J. Med.,* 282, 1231, 1970.

McCord, J. M. and Fridovich , I., The reduction of cytochrome c by milk xanthine oxidase, *J. Biol. Chem.,* 243, 5753, 1968.

McCord, J. M. and Fridovich, I., Superoxide dismutase: An enzyme function for erythrocuprein (hemocuprein), *J. Biol. Chem.,* 244, 6049, 1969.

McCord, J. M., Is iron sufficiency a risk factor in ischemic heart disease?, *Circulation,* 83, 1112, 1991.

McCord, J. M., Superoxide radical: controversies, contradictions, and paradoxes, *Proc. Soc. Exp. Biol. Med.,* 209, 112, 1995.

McCord, J. M., Iron, free radicals, and oxidative injury, *Semin. Hematol.,* 35, 5, 1998.

McKie, D., Wöhler's 'synthetic' urea and the rejection of vitalism: a chemical legend, *Nature,* 163, 608, 1944.

Miki, N., Kawabe, Y. and Kuriyama, K., Activation of cerebral guanylate cyclase by nitric oxide, *Biochem. Biophysic. Res. Commun.,* 75, 851, 1977.

186

Miller, G. J., High density lipoproteins and atherosclerosis, *Ann. Rev. Med.*, 31, 97., 1980.

Møller, P., Wallin, H. and Knudsen, L. E., Oxidative stress associated with exercise, psychological stress and lifestyle factors, *Chem. Biol. Int.*, 102, 17, 1996.

Nabel, E. G., Coronary heart disease in women - an ounce of prevention, *New Engl. J. Med.*, 343, 572, 2000.

Nakayama, T., Kaneko, M., Kodama, M. and Nagata, C., Cigarette smoke induces DNA single strand breaks in human cells, *Nature*, 314, 462, 1985.

Nelson, S. K., Bose, S. K. and McCord, J. M., The toxicity of high-dose superoxide dismutase suggests that superoxide can both initiate and terminate lipid peroxidation in the reperfused heart, *Free Radical Biol. Med.*, 16, 195, 1994.

Nowak, R., Mining treasures from junk DNA, *Science*, 263 (Feb. 4, 1994), 608.

Olusi, S. O., Ojutiku, O. O., Jessop, W., J. and Iboko, M. I., Plasma and white blood cell ascorbic acid concentrations in patients with bronchial asthma, *Clin. Chem. Acta*, 92, 161, 1979.

Palmer, R. M. J., Ferrige, A. G., and Moncada, S., Nitric oxide release accounts for the biological activity of endothelium-derived relaxing factor, *Nature*, 327, 524, 1987.

Palmer, R. M. J., Rees, D. D., Ashton, D. S. and Moncada, S., L-arginine is the physiological precursor for the formation of nitric oxide in endothelium-dependent relaxation, *Biochem. Biophys. Res. Commun.*, 153, 1251, 1988.

Piccirilli, J., RNA seeks its maker, *Nature*, 376 (Aug. 17, 1995), 548.

Pike, M. C. and Ross, R. K., Response, *J. Natl. Cancer Inst.,* 92, 1951, 2000.

Pitsikas, N. and Algeri, S., Deterioration of spatial and nonspatial reference and working memory in aged rats: protetive effect of life-long calorie restriction, *Neurobiol. Aging,* 13, 369, 1992.

Proudfoot, J. M., Croft, K. D., Puddey, 1. B. and Beilin, L. J., The role of copper reduction by α-tocopherol in low-density lipoprotein oxidation, *Free Radical Biol. Med,* 23, 720, 1997.

Prusiner, S. B., Novel proteinaceous infectious particles cause scrapie, *Science,* 216, 136, 1982.

Prusiner, S. B., Molecular biology of prion diseases, *Science,* 252, 1515, 1991.

Rahman, A., Sahabuddin, S. M., Hadi, S. M., Parish, J. H. and Ainley, K., Strand Scissions of DNA induced by quercentin and Cu(II): role of Cu(I) and oxygen free radicals in the reaction, Carcinogenesis, 10, 1833, 1989.

Richter, C., Park, J. W. and Ames, B. N., Normal oxidative damage to mitochondrial and nuclear DNA is extensive, *Proc. Natl. Acad. Sci. USA,* 85, 6465, 1988.

Rimm, E.B., Stampfer, M.J., Ascherio, A., Giovannucci, E., Colditz, G.A. and Willett, W.C., Vitamin E consumption and the risk of coronary heart disease in men, *N. Engl. J. Med.,* 328, 1450, 1993.

Roman-Franco, A. A., Non-enzymatic extramicrosomal bioactivation of chemical carcinogens by phagocytes: A proposed new pathway, *J. Theor. Biol.,* 97, 543, 1982.

Ross, R. K., Paganini-Hill, A., Wan, P. C. and Pike, M. C., Effect of hormone replacement therapy on breast cancer risk: estrogen versus estrogen plus progestin, *J. Natl. Cancer Inst.,* 92, 328, 2000.

188

Russell, G. A., Deuterium-isotope effects in the autoxidation of aralkyl hydrocarbons. Mechanism of the interaction of peroxyl radicals, *J. Am. Chem. Soc.*, 79, 3871, 1957.

Salonen, J. T., Salonen, R., Seppanen, K., Kantola, M., Suntioninen, S. and Korpela, H., Interactions of serum copper, selenium, and low density lipoprotein cholesterol in atherogenesis, *Brit. Med. J.*, 302, 756, 1991.

Salonen, J. T., Nyyssonen, K., Korpela, H., Tuomilehto, J., Seppanen, R. and Salonen, R., High stored iron levels are associated with excess risk of myocardial infarction in eastern Finnish men, *Circulation,* 86, 803, 1992.

Sawyer, D. T. and Valentine, J. S., How super is superoxide?, *Acc. Chem. Res.*, 14, 393, 1981.

Sbarra, A. J. and Karnovsky, M. L., The biochemical basis of phagocytosis. I. Metabolic changes during the ingestion of particles by polymorphonuclear leukocytes, *J. Biol. Chem.*, 234, 1355, 1959.

Schatzkin, A., Greenwald, P., Byar, D. P. and Clifford, C. K., The dietary fat-breast cancer hypothesis is alive, *JAMA,* 261, 3284, 1989.

Schimmel, P., and Alexander, R., All you need is RNA, *Science,* 281 (July 31, 1998), 658.

Schrödinger, E., An undulatory theory of the mechanics of atoms and molecules, *Physical Rev.*, 28, 1049, 1926.

Shigenaga, M. K., Gimeno, C. J. and Ames, B. N., Urinary 8-hydroxy-2'-deoxyguanosine as a biological marker of in vivo oxidative DNA damage, *Proc. Natl. Acad. Sci. USA*, 86, 9697, 1989.

Sies, H., Biochemistry of oxidative stress, *Angew. Chem. Int. Ed. Engl.*, 25, 1058, 1986.

Smith, K. C., Spontaneous mutagenesis: Experimental, genetic and other factors, *Mutation Res.*, 277, 139-162, 1992.

Sohal, R. S. and Weindruch, R., Oxidative stress, caloric restriction and aging, *Science,* 273, 59, 1996.

Solomon, G., Psychoneuroimmunology: Interaction between control nervous system and immune system, *J. Neurosci. Res.,* 18, 1, 1987.

Srivastava, L. M., Vasisht, S., Agarwal, D. P. and Goedde, H. W., Relation between alcohol intake, lipoproteins and coronary heart disease: the interest continues, *Alcohol and Alcoholism,* 29, 11, 1994.

Stadtman, E. R. and Oliver, C. N., Metal-catalyzed oxidation of proteins, *J. Biol. Chem.,* 266, 2005, 1991.

Stampfer, M. J., Hennekens, C.H., Manson, J.E., Colditz, G.A., Rosner, B.and Willett, W. C., Vitamin E consumption and the risk of coronary disease in women, *N. Engl. J. Med.*, 328, 1444, 1993.

Stefanick, M. L., Mackey, S., Sheehan, M., Ellsworth, N. Ellsworth, N., Haskell, W. L. and Wood, P. D., Effects of diet and exercise in men and postmenopausal women with low levels of HDL cholesterol and high levels of LDL cholesterol, *N. Engl. J. Med.,* 339, 12, 1988.

Steinbeck, M. J., Khan, A. U. and Karnovsky, M. J., Intracellular singlet oxygen generation by phagocytizing neutrophils in response to particles coated with a chemical trap, *J. Biol. Chem.,* 267, 13425, 1992.

Steinbeck, M. J., Khan, A. U. and Karnovsky M. J., Extracellular production of singlet oxygen by stimulated macrophages quantified using 9,10-diphenylanthracene and perylene in a polystyrene film, *J. Biol. Chem.,*268, 15649, 1993.

Steinberg, D., Antioxidant vitamins and coronary heart disease (editorial), *N. Engl. J. Med.*, 328, 1487, 1993.

190

Steinberg, D., Parthasarathy, S., Carew, T. E., Khoo, J. C. and Witztum, J. L., Beyond cholesterol: modifications of low density lipoprotein that increase its atherogenicity, *N. Engl. J. Med.,* 320, 915, 1989.

Steinbrecher, U. P., Zhang, H. and Lougheed, M., Role of oxidatively modified LDL in atherosclerosis, *Free Radical Biol. Med.,* 9, 155, 1990.

Stevens, R. G., Jones, D. Y., Micozzi, M. S. and Taylor, P. R., Body iron stores and risk of cancer, *N. Engl. J. Med.,* 319, 1047, 1988.

Stich, H. F., The beneficial and hazardous effects of some simple phenolic compounds, *Mutation Res.,* 259, 307, 1991.

Sullivan, J. L., Iron and the sex difference in heart disease risk, *Lancet,* 1 (8233), 1293, 1981.

Sullivan, J. L., The iron paradigm of ischemic heart disease, *Am. Heart J.,* 117, 1177, 1989.

Tannenbaum, A., The genesis and growth of tumors. II. Effects of caloric restriction per se, *Cancer Res.,* 2, 460, 1942.

Thaw, H. H., Brunk, U. T. and Collins, P. V., Influence of oxygen tension, prooxidants and antioxidants on the formation of lipid peroxidation products (lipofuscin) in individual cultivated human glial cells, *Mech. Aging Develop.,* 24, 211, 1984.

Timoféeff-Ressovsky, N. W., Zimmer, K. G. and Delbrück, M., Über die Natur der Genmutation und der Genstruktur, *Nachr. Ges. Wiss. Göttingen,* 6 NF (13), 190, 1935.

Toshniwal, P. K. and Zarling, E. J., Evidence for increased lipid peroxidation in multiple sclerosis, *Neurochem. Res.,* 17, 205, 1992.

Toth, K. M., Berger, E. M., Beehler, C. J. and Repine, J. E., Erythrocytes from cigarette smokers contain more glutathione and catalase and protect endothelial cells from hydrogen peroxide better than do erythrocytes from nonsmokers, *Am. Rev. Resp. Dis.,* 134, 281, 1986.

Viguie, C. A., Frei, B., Shigenaga, M. K., Ames, B. N., Packer, L. and Brooks, G. A., Antioxidant status and indexes of oxidative stress during consecutive days of exercise, *J. Appl. Physiol.,* 75, 566, 1993.

Weindruch, R., Walford, R. L., Fligiel, S and Guthrie, D., The retardation of aging in mice by dietry restriction restriction: longevity, cancer, immunity and lifetime energy intake, *J. Nutr.,* 116, 641, 1986.

Weitz, Z. W., Birnbaum, A. J. and Skosey, J. L., High breath pentane concentrations during acute myocardial infarction, *Lancet,* 337, 933, 1991.

Weitzman, S. A. and Gordon, L. I., Inflammation and cancer: role of phagocyte-generated oxidants in carcinogenesis, *Blood,* 76, 655, 1990.

Willett, W. C. et al. Intake of *trans* fatty acids and risk of coronary heart disease among women, *Lancet,* 341, 581, 1993.

Yakes, F. M. and Van Houten, B., Mitochondrial DNA damage is more extensive and persists longer than nuclear DNA damage in human cells following oxidative stress, *Proc. Natl. Acad. Sci. USA,* 94, 514, 1997.

Yong, L. C., Brown, C. C., Schatzkin, A., Dresser, C. M., Slesinski, M. J., Cox, C. S. and Taylor, P. R., Intake of vitamins E, C. and A and risk of lung cancer. The NHANES I epidemiologic followup study. First National Health and Nutrition Examination Survey, *Am. J. Epidemiol.,* 146, 231, 1997.

Young, A. C. et al., The effect of atherosclerosis on the vasomotor response of coronary arteries to mental stress, *N. Engl. J. Med.,* 325, 1551, 1991.

Youngman, L. D., Park, J. Y. and Ames, B. N., Protein oxidation associated with aging is reduced by dietary restriction of protein or calories, *Proc. Natl. Acad. Sci., USA,* 89, 9112, 1992.

Youngman, L. D. and Campbell, T. C., Inhibition of aflatoxin B_1-induced gamma-glutamyltranspeptidase positive (GGT⁺) hepatic preneoplastic foci and tumors by low protein diets: evidence that altered GGT⁺ foci indicate neoplastic potential, *Carcinogenesis,* 13, 1607, 1992.

Yu, S. Y., Chu, Y. J., Gong, X. L. and Hou, C., Regional variation of cancer mortality incidence and its relation to selenium level in China, *Biol. Trace Elem. Res.*, 7, 21, 1985.

Ziegler, R. G., Hoover, R. N .Pike, M. C. et al., Migration patterns and breast cancer risk in Asian-American women, *J. Natl. Cancer Inst.,* 85, 1819, 1993.

Subject Index

Appendix

Table I   L-Amino Acids Found in Proteins

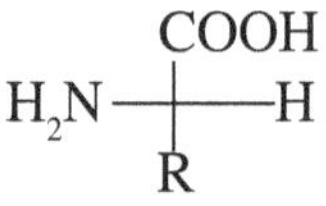

$$H_2N - \overset{\displaystyle COOH}{\underset{\displaystyle R}{\vert\!\!-\!\!\vert}} - H$$

STRUCTURE OF R	NAME

R group is neutral

-H   Glycine

$-CH_3$ Alanine

$-CH(CH_3)_2$ Valine[e]

$-CH_2CH(CH_3)_2$ Leucine[e]

$-\underset{\displaystyle CH_3}{CHCH_2CH_3}$ Isoleucine[e]

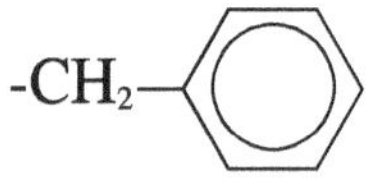

$-CH_2-\bigcirc$ Phenylalanine[e]

$-CH_2CONH_2$ Asparagine

$-CH_2CH_2CONH_2$ Glutamine

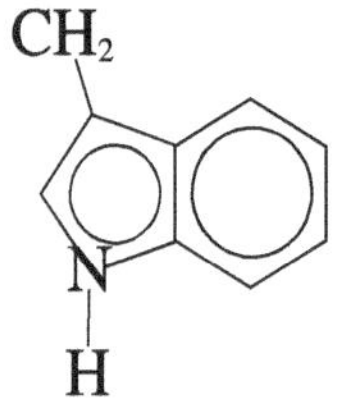

Tryptophan[e]

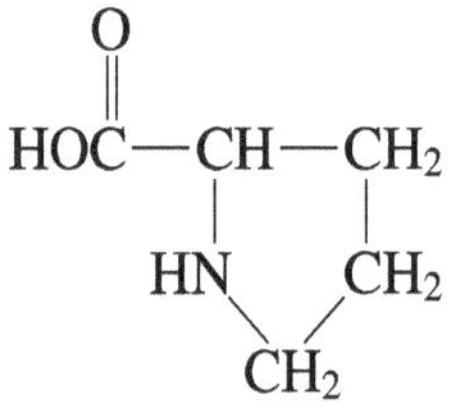

$$HO\overset{O}{\overset{\|}{C}} - CH - CH_2$$

Proline
(complete structure)

Table I, continued

STRUCTURE OF R	NAME
R contains an $-OH$ group	
$-CH_2OH$	Serine
$-CHOH$ $\quad | $ $\quad CH_3$	Threonine[e]
 $-CH_2-\bigcirc-OH$	Tyrosine
R contains sulfur	
$-CH_2SH$	Cysteine
$-CH_2CH_2SCH_3$	Methionine[e]
R contains a carboxyl group	
$-CH_2COOH$	Aspartic acid
$-CH_2CH_2COOH$	Glutamic acid
R contains a basic amino group	
$-CH_2CH_2CH_2CH_2NH_2$	Lysine[e]
$-CH_2CH_2CH_2NH_2-\overset{\overset{\displaystyle NH}{\|\|}}{C}-NH_2$	Arginine
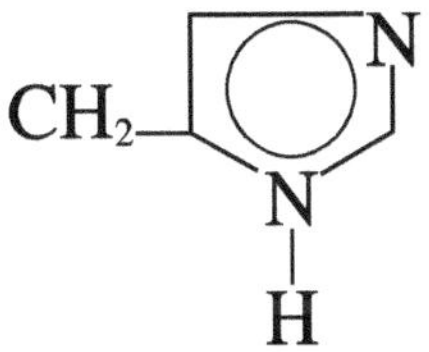 $CH_2-\bigcirc$ (imidazole ring with N, N-H)	Histidine

e = essential amino acids

Table II, Basic Building Blocks

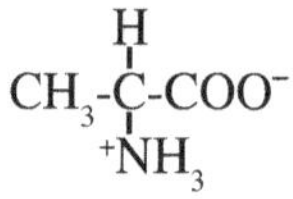

$$CH_3-\overset{\overset{\displaystyle H}{|}}{\underset{\underset{\displaystyle +NH_3}{|}}{C}}-COO^-$$

alanine zwitter ion

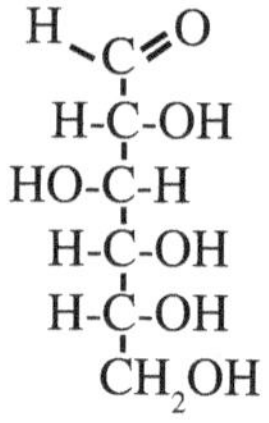

$$CH_3-\overset{\overset{\displaystyle H}{|}}{\underset{\underset{\displaystyle NH_2}{|}}{C}}*-COOH$$

alanine

$$CH_3-(CH_2)_{14}-COOH$$

palmitic acid

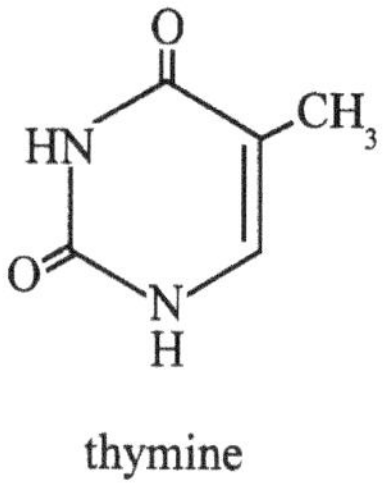

$$\begin{array}{c}
H_{\diagdown}C{\diagup}^O \\
H-C-OH \\
HO-C-H \\
H-C-OH \\
H-C-OH \\
CH_2OH
\end{array}$$

D-glucose

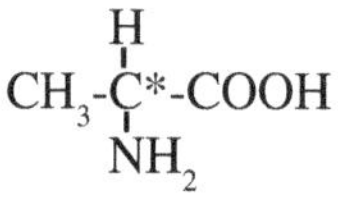

$$CH_3-(CH_2)_7-CH{=}CH-(CH_2)_7-COOH$$

oleic acid

thymine

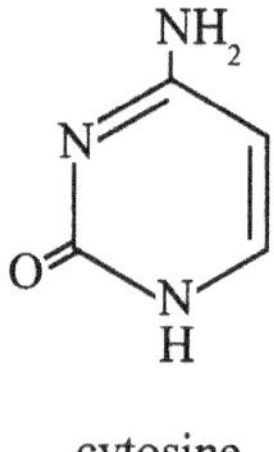

cytosine

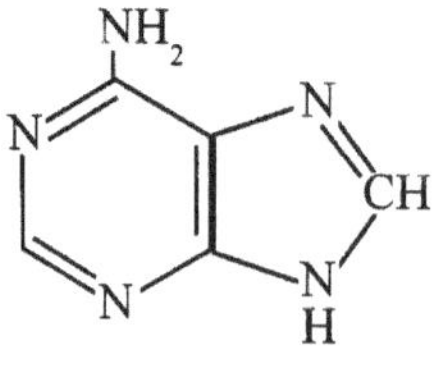

adenine

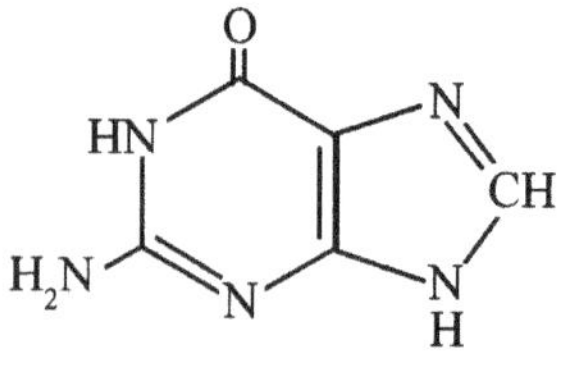

guanine

Table III, DNA Strand and Base Pairs

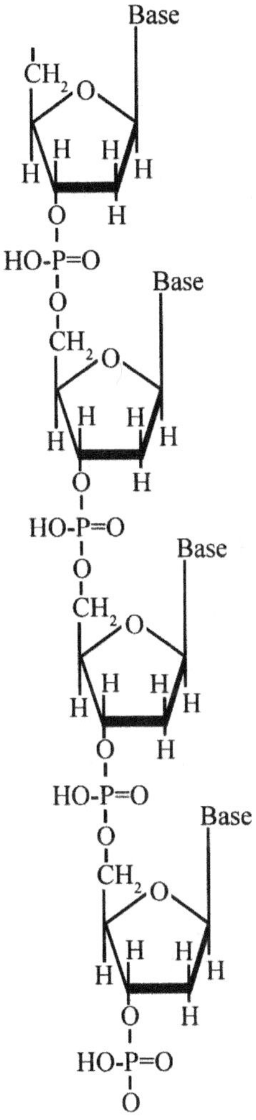

Single strand of DNA

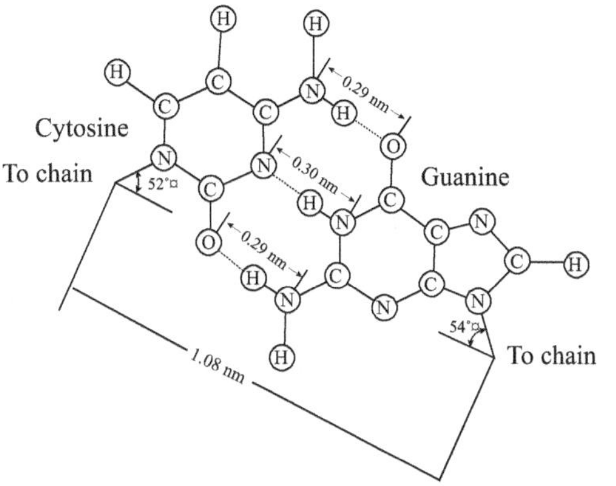

Guanine-cytosyne pair

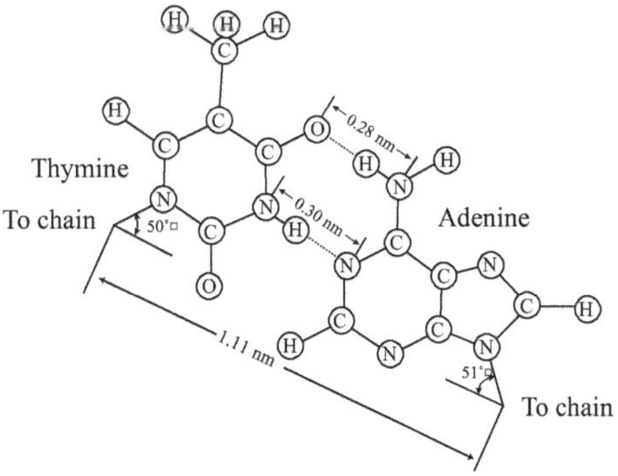

Adenine-thymine pair

PERIODIC TABLE OF THE ELEMENTS

Group / Row	I	II						Transition elements						III	IV	V	VI	VII	0
1	H 1																		He 2
2	Li 3	Be 4												B 5	C 6	N 7	O 8	F 9	Ne 10
3	Na 11	Mg 12												Ai 13	Si 14	P 15	S 16	Cl 17	Ar 18
4	K 19	Ca 20	Sc 21	Ti 22	V 23	Cr 24	Mh 25	Fe 26	Co 27	Ni 28	Cu 29	Zn 30	Ga 31	Ge 32	As 33	Se 34	Br 35	Kr 36
5	Rb 37	Sr 38	Y 39	Zr 40	Nb 41	Mo 42	Tc 43	Ru 44	Rh 45	Pd 46	Ag 47	Cd 48	In 49	Sn 50	Sb 51	Te 52	I 53	Xe 54
6	Cs 55	Ba 56	*La 57-71	Hf 72	Ta 73	W 74	Re 75	Os 76	Ir 77	Pt 78	Au 79	Hg 80	Ti 81	Pb 82	Bi 83	Po 84	At 85	Rn 86
7	Fr 87	Ra 88	†Ac 89-103																

*Lanthanide; †Actinide

www.ingramcontent.com/pod-product-compliance
Lightning Source LLC
Chambersburg PA
CBHW060459290526
45791CB00001B/193